PRE-CONTRACT PRACTICE

FOR THE BUILDING TEAM

'For which of you, intending to build a tower, sitteth not down first, and counteth the cost, whether he have sufficient to finish it.'

ST LUKE XIV. 28

PRE-CONTRACT PRACTICE

FOR THE BUILDING TEAM

EIGHTH EDITION

THE AQUA GROUP

With sketches by
Brian Bagnall

OXFORD

BLACKWELL SCIENTIFIC PUBLICATIONS

LONDON EDINBURGH BOSTON
MELBOURNE PARIS BERLIN VIENNA

Blackwell Scientific Publications
Editorial Offices:
Osney Mead, Oxford OX2 0EL
25 John Street, London WC1N 2BL
23 Ainslie Place, Edinburgh EH3 6AJ
3 Cambridge Center, Cambridge
   Massachusetts 02142, USA
54 University Street, Carlton
   Victoria 3053, Australia

Other Editorial Offices:
Librarie Arnette SA
2, rue Casimir-Delavigne
75006 Paris
France

Blackwell Wissenschafts-Verlag
Meinekestrasse 4
D-1000 Berlin 15
Germany

Blackwell MZV
Feldgasse 13
A-1238 Wien
Austria

First edition published by Crosby,
   Lockwood & Son Ltd 1960
Second edition 1964
Third edition 1967
Fourth edition 1971
Fifth edition published by Crosby
   Lockwood Staples 1974
Sixth edition published by
   Granada Publishing Ltd 1980
Seventh edition published by Collins
   Professional and Technical Books 1986
Reprinted by BSP Professional Books 1990
Eighth edition published by Blackwell
   Scientific Publications 1992
Set by CG Graphic Services, Aylesbury, Bucks
Printed and bound in Great Britain by
Hartnolls Ltd, Bodmin, Cornwall

DISTRIBUTORS

Marston Book Services Ltd
PO Box 87
Oxford OX2 0DT
(Orders: Tel: 0865 79115
        Fax: 0865 791927
        Telex: 837515

USA
Blackwell Scientific Publications, Inc
3 Cambridge Center
Cambridge, MA 02142
(Orders: Tel: 800 759-6102
                617 225-0401)

Canada
Oxford University Press
70 Wynford Drive
Don Mills
Ontario M3C 1J9
(Orders: Tel: 416 441- 2941)

Australia
Blackwell Scientific Publications
(Australia) Pty Ltd
54 University Street
Carlton, Victoria 3053
(Orders: Tel: 03 347-0300)

British Library
Cataloguing in Publication Data
A Catalogue record for this book is
available from the British Library

ISBN 0-632-02817-3

Library of Congress
Cataloging in Publication Data

Pre-contract practice for the building team/the
Aqua Group; with sketches by Brian Bagnall.
—8th ed.
   p. cm.
   Rev. ed. of: Pre-contract practice for
architects and quantity surveyors. 3rd ed. rev.
and enl. 1967.
   Includes index.
   ISBN 0–632–02817–3
   1. Building—Economic aspects.
2. Building—Specifications.
3. Construction industry—Management.
I. Aqua Group.   II. Pre-contract practice for
architects and quantity surveyors.
TH435.P715   1992
692—dc20   92–12075
                                      CIP

CONTENTS

Authors' Note to the First Edition

At a meeting arranged by the Junior Organisation Quantity Surveyors' Committee of The Royal Institution of Chartered Surveyors, held at the Talbot Restaurant, London Wall, a number of architects were present, and following a talk by a well known builder on 'The Preparation of a Builder's Estimate', there was a lively discussion from which it appeared that many of the troubles which arose in building were due to inadequate pre-contract preparation.

It was suggested that if some of the younger members of our two professions felt that was the case, they should get together and do something about it.

In answer to that challenge a group of us dined together and talked about our work. From that beginning, *Pre-Contract Practice for Architects and Quantity Surveyors* has emerged.

For convenience, we adopted the name 'Aqua'. There is no significance in the choice of the word and, indeed, the reasons for choosing it are now lost in obscurity. For us, however, it will always be a reminder of many hours of discussion and hard work, a full measure of good humour, and much learned by members of two professions about each other's problems.

The sketches were drawn by Brian Bagnall.

PETER JOHNSON, FRICS, FIArb (Chairman)
H. E. D. ADAMSON, FRIBA
HARRY L. AGER, ARICS
BRIAN BAGNALL, BArch (L'pool)
A. T. BRETT-JONES, ARICS
F. S. JOHNSTONE, ARICS
JOHN KEMP, ARIBA
A. G. NISBET, BA (Arch), FRIBA
C. A. ROGER NORTON, AA Dipl (Hons), FRIBA
A. J. OAKES, FRICS, FIArb

August 1959

Introduction

The purpose of the Aqua Group has evolved over the years and we have come to see it as setting down in clear, concise and practical terms the principles of good practice in the work of the architect, quantity surveyor and other members of the building team.

All the Aqua books assume the use of the current edition for the time being of the JCT Standard Form of Building Contract and the selection of a contractor by competitive tendering.

In the thirty years since the first edition of this book was published there have been considerable changes in practice and procedure. The composition of the Aqua Group has also changed. Now only two of the original members of the group continue actively to contribute but the newer members have not lost sight of the original objectives.

The first book, *Tenders and Contracts*, deals with the various alternative methods of procurement, tendering procedures and contractual arrangements and the circumstances in which they might be adopted in order that consideration may be given to the most advantageous method for a particular project.

Pre-Contract Practice is the second of our trilogy of books on good practice and covers the important period between inception and the invitation to tender.

The period from receipt of tenders to the placing of the contract, through the post-contract period to the settlement of the final account, is covered in *Contract Administration*.

Such is the pace of change in building procedures and contracts that all three books are in a continuous state of review.

In preparing this, the eighth edition of *Pre-Contract Practice*, the Group has recognized that the pace of change has led to an increasing trend over recent years towards alternative methods of building procurement which have affected procedures. The first two chapters discuss those factors which should be considered before an employer can be properly advised. With the increasing complexity of projects, the book describes in greater detail in Chapter 2 the essential interrelationship between the methods of procurement and the development of the brief, from initial statement of intent to final design stage, emphasizing the linking between development of the design brief and the design process.

The way one sets about determining the right procedure depends on the method of procurement and type of contract chosen and these chapters should be read together with *Tenders and Contracts*. The succeeding chapters have been extensively revised but follow the pattern of earlier editions and concentrate on those procedures which should ensure good practice on a traditional building project.

Whichever procedures are adopted certain fundamental principles should always be followed.

First, it is essential that the policy to be followed is decided as near to the inception of the project as is practicable so that the pre-contract programme and procedure can be prepared accordingly.

Second, the method of appointing the contractor, whether by single- or two-stage competitive tender, negotiation or nomination, should be decided according to the requirements of the particular case.

Third, the form of contract selected should be suitable for the procedure adopted.

Finally, the design and specification of the work must always be sufficiently ahead of construction to enable the contractor to plan the work properly, obtain materials and carry out the work in an efficient and economic way. These principles should be adhered to throughout.

As employers look increasingly for an assurance of quality from their professional advisers so it becomes important for practices not only to provide the level of quality expected but to be seen to do so. The application of the management process of quality assurance is discussed in Chapter 2.

The concept of co-ordinated project information (CPI) referred to in Chapter 4 introduced a structured method for the cross-referencing of data between all facets of the information process, with a view to ensuring that bills of quantities are produced with the appropriate drawings and specifications. The link between CPI and the seventh edition of the Standard Method of Measurement (SMM7, published in 1988) is referred to in Chapter 6.

Most offices now use computers, from the word processor to more sophisticated hardware and software, which assist in the preparation and dissemination of information throughout the building team. The current storage facility of a simple personal computer permits the preparation, storage and retrieval of specifications, schedules, and other standard documentation, combining consistency of information with a significant saving in time. With the advent of more advanced software packages the use of computer aided design (CAD) can be advantageous and this is referred to in Chapter 4.

The implications of the changes in the method of measurement from trade sections to work sections within bills of quantities prepared in accordance with SMM7 are discussed in Chapter 6.

The somewhat complex procedures for the nomination of sub-contractors as required by JCT 80 have been modified by Amendment 10 and this is covered in Chapter 7.

We believe that all those engaged in the business of building, from student to practitioner,   must have a proper understanding of good practice in relation to a traditional building project before the various alternative options can be explored. After the more general discussion in the opening chapters, recommendations are based on the premise of dealing with a typical and reasonably normal project, for which a recognized competitive tendering procedure is adopted and for which the JCT Standard Form of Building Contract, 1980 Edition, with Quantities is used. Where appropriate, reference is made to the alternative procedures discussed in the earlier chapters.

This book, together with *Tenders and Contracts for Building* and *Contract Administration*, will give comprehensive guidance for those involved in the management of building projects. If these recommendations are followed then a good standard of practice will be established. Modifications to procedure can be made to suit changing circumstances and conditions without affecting the high standard of work which the professional team have a duty to provide.

As always the Group is grateful to Brian Bagnall for his delightful cartoons introducing a humorous tone.

The Aqua Group

QUENTIN PICKARD, BA, RIBA (Chairman)
BRIAN BAGNALL, BArch (L'pool)
HELEN DALLAS, BA, Dip Arch, RIBA
JEREMY NEWTON, ARICS
JOHN OAKES, FRICS, FCIArb
RICHARD OAKES, BSc, FRICS
GEOFFREY POOLE, FRIBA, ACIArb
GEOFF QUAIFE, ARICS
JOHN TOWNSEND, FRICS, ACIArb
JOHN WILLCOCK, Dip Arch, RIBA
JAMES WILLIAMS, DA (Edin), FRIBA

To design ... down to the last ... doorknob

Chapter 1

Assessing the Needs

The Aim

Success in building requires that a complicated series of interactions be completed in a logical and pre-determined sequence. The employer will then get the building he wants, at an acceptable price, in the required time. This book aims to set down the principles of good practice up to the point of inviting tenders which when applied to the pre-contract process should ensure that this common goal is achieved.

The first two chapters present a discussion of those factors which need to be addressed in developing a scheme from the earliest involvement of a consultant through to the decision by the employer to proceed with the preparation of the appropriate information necessary to obtain tenders. Much preliminary work may have already been undertaken by the employer in establishing his need for a new building, or in assessing the market for speculative buildings. Alternative procurement routes and their effect on the appointment and responsibilities of consultants are discussed. From Chapter 3 onwards our recommendations for ensuring good practice generally assume a traditional building project to be let under the JCT 80 Standard Form of Building Contract.

The Structure

The successful development of any building project requires the acceptance by all parties of an underlying structure or framework of operation. For a traditional building project the Outline Plan of Work published by the RIBA, and reproduced here, describes an excellent framework on which to base the structure of the pre-contract process. It also provides a useful checklist for key activities.

As work proceeds, it is prudent to reconsider earlier decisions to ensure that the developing design continues to satisfy the employer's overall criteria.

Generally, the procedures recommended in this book follow the Outline Plan of Work. The plan, however, can be modified to suit those

1

Outline plan of work
Plan of work diagram 1

Stage	Purpose of work and decisions to be reached	Tasks to be done	People directly involved	Usual terminology
A. Interception	To prepare general outline of requirements and plan future action.	Set up client organization for briefing. Consider requirements, appoint architect.	All client interests, architect.	**Briefing**
B. Feasibility	To provide the client with an appraisal and recommendation in order that he may determine the form in which the project is to proceed, ensuring that it is feasible, functionally, technically and financially.	Carry out studies of user requirements, site conditions, planning, design, and costs, etc., as necessary to reach decisions.	Clients' representatives, architects, engineers, and QS according to nature of project.	
C. Outline Proposals	To determine general approach to layout, design and construction in order to obtain authoritative approval of the client on the outline proposals and accompanying report.	Develop the brief further. Carry out studies on user requirements, technical problems, planning, design and costs, as necessary to reach decisions.	All client interests, architects, engineers, QS and specialists as required.	**Sketch Plans**
D. Scheme design	To complete the brief and decide on particular proposals, including planning arrangement appearance, constructional method, outline specification, and costs, and to obtain all approvals.	Final development of the brief, full design of the project by architect, preliminary design by engineers, preparation of cost plan and full explanatory report. Submission of proposals for all approvals	All client interests, architects, engineers, QS and specialists and all statutory and other approving authorities.	

Brief should not be modified after this point.

			Working drawings
E. Detail design	To obtain final decision on every matter related to design, specification, construction and cost.	Full design of every part and component of the building by collaboration of all concerned. Complete cost checking of designs.	Architects, QS, engineers and specialists, contractor (if appointed).

Any further change in location, size, shape, or cost after this time will result in abortive work.

F. Production information	To prepare production information and make final detailed decisions to carry out work.	Preparation of final production information, i.e. drawings, schedules and specification.	Architects, engineers and specialists, contractor (if appointed).
G. Bills of quantities	To prepare and complete all information and arrangements for obtaining tender.	Preparation of bills of quantities and tender documents.	Architects, QS, contractor (if appointed).
H. Tender action	Action as recommended in NJCC *Code of Procedure for Single Stage Selective Tendering* 1977, April 1989 Revision.	Action as recommended in NJCC *Code of Procedure for Single Stage Selective Tendering* 1977, April 1989 Revision.	Architects, QS, engineers, contractor, client.

Reproduced by permission of RIBA Publications Ltd.
Note: On small jobs without quantities Stage G will be omitted and Stages E and F may be combined.

pre-contract procedures required by alternative methods of building procurement.

The Brief

The initial brief from the employer will often be little more than a statement of intent; it may be no more than a telephone call. At this stage there is unlikely to be any formal appointment. After the initial euphoria surrounding 'getting a job' the consultant will quickly realize that far more information is required from both the potential employer, and from within the consultant's own organization.

From the employer, the consultant will typically require details of:

- The status of the employer; owner occupier, developer, contractor, purchaser.
- The purpose behind the employer's desire to build; requirement of accommodation, profit, or investment (see Chapter 2).
- An assessment of the employer's financial resources to complete the project.
- Any technical skills the employer can contribute to the project.
- Is the consultant being interviewed for the job in competition with other professionals?
- Does the employer have a site? Details of any preliminary discussions that may have taken place with planning authorities or other statutory authorities.
- An indication of timescale.
- Details of any preliminary work the employer has undertaken to establish the need for, or market for, the new building.

From within his own organization the consultant will need to monitor:

- Current workload: this will help to determine how much the consultant actually wants the appointment.
- Is the consultant qualified to undertake the project?
- Availability of manpower: when and for how long, or will new staff need to be recruited?
- Availability of specialist technical skills.
- Acceptable level of risk: is there any element of speculative work or project development for which no fee will be paid if the job does not proceed beyond a certain stage? The consultant will often be expected to undertake a degree of speculative work in assisting in the preparation of a site bid where the prospective employer is a developer.

- The effect of different procurement routes on the consultant's responsibilities and liabilities.

The Initial Programme

One of the first questions asked by the employer will be 'How quickly can I have my building?' At this very early stage it will be impossible for the consultant to set a hard and fast pre-contract programme. However, a simple barchart setting a timescale for the various key activities will provide both the employer and the design team with a useful framework (Example A). The chart would need to identify clearly any activities where the timescale for making decisions is beyond the control of the design team: for example, in obtaining planning approvals.

At this stage the employer can only rely on the professional judgement of his advisers on the likely implications of any difficulties. The charts contained within this book have assumed a straightforward non-

The likely implications of any difficulties

EXAMPLE A: PRE-CONTRACT PROGRAMME

general notes
SPECIFIC DATES SHEWN IN CIRCLES
DOUBLE LINES INDICATE KEY EVENTS

revisions

project/client
SHOPS AND OFFICES
NEW BRIDGE STREET
BORCHESTER

drawing title
pre-contract programme

number 456/21

rev
scale   date   drawn

Reed & Seymore architects
12 The Broadway
Borchester BC4 2NW

Chart rows (ITEM):
- FEASIBILITY STUDY
- OUTLINE PROPOSALS (SKETCH PLANS)
- PLANNING APPLICATION (IN OUTLINE)
- SCHEME DESIGN
- DETAIL DESIGN
- CONSULTANTS DRAWINGS ETC.
- PRODUCTION INFORMATION
- SUB-CONTRACTORS ESTIMATES ETC.
- STATUTORY APPROVALS
- BILLS OF QUANTITIES
- PRINTING OF BILLS TENDER DOCUMENTATION
- TENDER ACTION
- ANALYSE/REPORT ON TENDERS
- PRE-CONTRACT PLANNING

Week Ending / Week Number across months: AUG, SEPTEMBER, OCTOBER, NOVEMBER, DECEMBER, JANUARY, FEBRUARY, MARCH, APRIL, MAY, JUNE

Key event annotations:
- CLIENT APPROVAL
- BRIEF SHOULD NOT BE MODIFIED AFTER THIS POINT
- ANY FURTHER CHANGES WILL RESULT IN ABORTIVE WORK
- INFORMATION & SPEC. NOTES
- COMPLETION OF PRODUCTION
- DOCUMENTS SENT OUT
- TENDERS DUE FOR RECEIPT
- CONTRACT SIGNING
- SITE WORKS COMMENCE (27)

controversial project which should not encounter major delays.

In setting the initial programme it should be recognized that it will be unlikely that the procurement route (drawings and bills of quantities, design and build, management contract) will have been established. Therefore, any assumptions which have been made in the absence of firm decisions should be confirmed during the development of the brief, and adjustments made to the programme accordingly.

The assessment of the resources required for the project and the agreement to an initial programme will assist the consultant in advising on the development brief and the preparation of the detailed pre-contract programme.

The Appointment

Traditionally an employer's first appointment would normally be his architect, who would then assume the role of project leader. The architect would be expected to advise on the appointment of other consultants, although appointments should always be made direct with the employer. This avoids contractual liability on the part of the architect for the consultants' professional performance, and the payment of their fees. Increasingly, this traditional arrangement is being modified to suit alternative methods of building procurement which may result in a different project leader.

Following the negotiations between the Office of Fair Trading and the professional organizations to abolish mandatory fee scales, many employers now invite fee bids from consultants relating to a particular level of service. When few details of the scheme are certain, consultants have come to realize that preparing realistic fee bids at an early stage in a project is both difficult and financially hazardous. Equally, the sensible evaluations of competing bids is correspondingly difficult. Just as in the submission of an unrealistically low building tender, the employer should be aware of the unrealistically low fee bid, and question the consultant's ability satisfactorily to undertake the work. Unfortunately, in any competitive tender/fee bid situation, it is unusual for anything other than the lowest price to be accepted. It follows that the fullest possible documentation should be available to enable the consultant to prepare a realistic fee bid.

Appointment Documents

All consultants should establish in writing their terms of appointment, covering:

- Who the employer is;
- The services that are to be provided;
- The conditions of appointment;
- How fees are to be calculated, and when they are to be paid;
- Terms that will apply should the project change or not proceed.

Most professional organizations now have their own contract of appointment for execution between the employer and the consultant. Typically, these documents will specify the services to be provided and the fee to be paid for carrying out these services. Guidance is given on appropriate fee scales covering a range of building types, which gives assistance both to the consultant in preparing the fee proposal and to the client in assessing whether it is realistic.

In addition to the normal appointment documents consultants are increasingly being required to enter into collateral warranties. A collateral warranty provides an agreement that is an adjunct to another, or principal agreement (the appointment). They are commonly used to bind a consultant into contract for a specific period of time with a third party, frequently the funding institution, where no contract would otherwise exist.

Chapter 2

From Brief to Tender:
A Pre-Contract Procedure

Initial Brief

The initial brief from the employer may be little more than a statement of intent as was mentioned in Chapter 1. However, for the project to be successful, the employer's brief must be developed in detail so that due consideration can be given to each aspect.

The design brief and the design process develop in cycles, with progress on one aspect creating a need for further thought on and consideration of other matters. The more complicated the building the more important it is that the brief is developed in a systematic way and as early as possible. Full development, certainly on major schemes, is a team affair requiring the preparation of feasibility studies, cost reports and consultations with planning and other relevant authorities.

Developing the Brief

Numerous factors need to be taken into account when developing the brief. Some of the main ones are:

- *Site* – Should be evaluated to explore factors such as access limitations, site topography and soil suitability (using soil tests, trial and bore holes). A detailed survey will be required of the site and/or buildings and this should provide critical information such as dimensions, boundaries, levels, existing services, watercourses, access, trees and conditions of buildings.
- *Costs* – Firm details of the employer's budget are required by the design team.
- *Finance* – Funding arrangements should be set in place giving due regard to cash flow constraints, overall capital and projected running costs.
- *Timescale* – A programme should be assessed and the appropriateness of the desired commencement and completion dates established in the context of the practicalities of the situation and other buildings the employer may be planning.
- *Operational factors* – Determining the activities the building must

9

accommodate and the practical factors which will govern the layout, relationships and priorities of the various elements required in the scheme. For example, an assessment should be made of the need to isolate certain aspects of plant and machinery in order to create acceptable noise and environmental conditions in the remainder of the building. Additionally, the degree of flexibility in the designed space and the future adaptability of the building should be considered.

- *Market* – Speculative buildings need examination of the market potential, timing and selling strategies.
- *Employer attitude* – Must the building be designed to a strict corporate standard or, if further finance became available, would the scheme be varied?
- *Environment* – Type and quality of the internal and external environment desired.

Feasibility Stage

Developing the design and costings will lead to a feasibility report, advising the employer whether the project is feasible functionally, technically and financially.

Sketch Scheme

The design process usually commences with the sketch scheme. Initially this involves the designer or architect putting to paper their preliminary responses to the brief. Depending on the nature of the project, outline plans may demonstrate the extent of accommodation that can be achieved: for example, the number of new houses on a particular site, the area of office space against a nett:gross ratio, or the number of shopping units with their associated parking requirements. In conjunction with an outline specification of type of structure, construction and level of finish and servicing, this may provide sufficient information for an employer to move on to the cost plan stage.

However, if the emphasis is on design quality as in a conservation area or an important site or sensitive extensions to buildings, then the sketch scheme will concentrate equally on the visual form, style, proportion and materials, demonstrated in preliminary elevations or sketch perspectives. Whatever type of drawings or information is produced at this stage, the presentation should be readily understandable by the parties examining the scheme.

Unless the employer has dealt with the issue, the design team should now advise on planning legislation where relevant. Firstly, does the proposal require consent? If not, it is wise to obtain confirmation of this in case of later dispute. Assuming consent is required, the need to obtain an outline planning consent prior to detail should be discussed. This usually becomes relevant when a principle needs to be established: for example a major change of use, or the development of a site not designated under a local authority plan, or the development of a site over and above local density levels. Unless in a sensitive area, only basic plans or Ordnance Survey map extracts are required, thereby limiting the financial commitment of the employer.

Costs

At the start of the sketch scheme the quantity surveyor may comment on the projected capital costs by preparing a preliminary approximate estimate – a simple cost per square metre calculation or outline unit cost. As the scheme develops it will be possible for the initial cost estimate to be amended or confirmed. This establishes an overall budget for the scheme and provides a cost control document. The design and cost of each element can be monitored and prevented from unintentionally becoming unduly expensive or overdesigned in comparison with other elements.

The employer is now in possession of an outline scheme which meets the basic parameters of his brief together with a cost plan based on set levels of quality. A decision whether or not to proceed can now be made; beyond this point a commitment to advance the scheme to the next stage will engender significant financial outlay.

Procurement

The method of choosing the contractor to construct the building must be determined, as this may dictate the manner in which the detailed design is progressed. For example, on a traditional procurement route the design team will develop the design whereas with design and build procurement the design team may only prepare a design brief and the design itself will be completed by the contractor. Final acceptance of the procurement method needs to be made at this stage as all the key elements of the brief will have been established.

Each procurement option will have a different time, cost, quality scenario (the 'procurement triangle'). These three elements may be held in a particular balance as the circumstances dictate that one element must take precedence over the others. However, altering any one element will

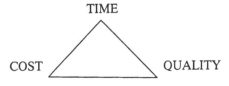

TIME

COST   QUALITY

Procurement triangle

have an effect on the others.

One interpretation of the time, cost and quality priorities which may be derived from three different procurement routes is given in Example B. It must be stressed that the particular circumstances of each project may result in different conclusions as to the procurement route which best satisfies the priorities of time, cost and quality. For example, a high level of provisional sums within a contract could undermine the cost certainty afforded by a particular procurement route. Additionally, it should be noted that for ease of presentation time allocation for planning application approval has been omitted from Example B. This element in itself can dictate or influence the choice of the procurement route.

For full reference to procurement routes and the forms of contract which accompany them, refer to the Aqua book *Tenders and Contracts*.

It is advisable to report procurement options to the employer, recommending the most appropriate route, detailing how the priorities are protected and outlining implications.

EXAMPLE B: PROCUREMENT OPTIONS

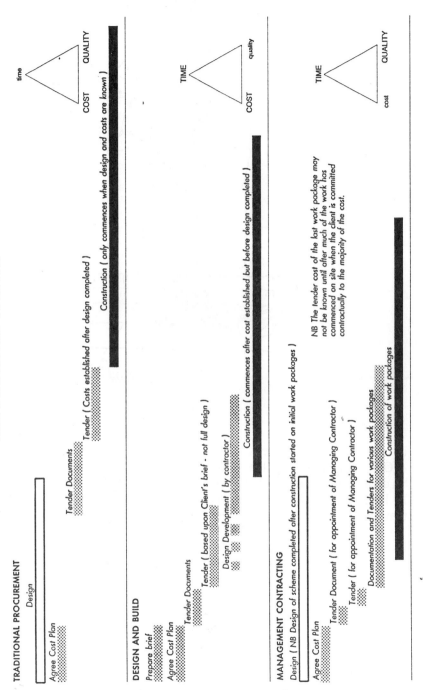

PROCUREMENT OPTIONS

TRADITIONAL PROCUREMENT

Design

Agree Cost Plan

Tender Documents

Tender (Costs established after design completed)

Construction (only commences when design and costs are known)

time / COST / QUALITY

DESIGN AND BUILD

Prepare brief

Agree Cost Plan

Tender Documents

Tender (based upon Client's brief - not full design)

Design Development (by contractor)

Construction (commences after cost established but before design completed)

TIME / COST / quality

MANAGEMENT CONTRACTING

Design (NB Design of scheme completed after construction started on initial work packages)

Agree Cost Plan

Tender Document (for appointment of Managing Contractor)

Tender (for appointment of Managing Contractor)

Documentation and Tenders for various work packages

Construction of work packages

NB The tender cost of the last work package may not be known until after much of the work has commenced on site when the client is committed contractually to the majority of the cost.

TIME / cost / QUALITY

Only when the method of procurement is agreed can the pre-contract programme be finalized. This also represents the point at which the specific service agreements and fee proposals can be confirmed for the design team and individual roles defined.

Detail Design

The development of the sketch scheme into a workable solution will produce the detail design to the employer's brief and allow it to be submitted for detailed planning consent. The onus is on the design team to maintain the priorities established within the employer's brief and to extract and agree any further information required from the employer as the scheme progresses. Whatever the scale of the project the design team have a responsibility to use their skills to:

- Produce a good solution to meet their employer's brief;
- Consider end or future users if different from the employer;
- Design to meet the anticipated lifespan;
- Design to avoid early deterioration or costly maintenance;
- Consider an environmental responsibility to the public who may visit or pass the building;
- Co-ordinate structure, finishings and services into one complete design.

During this development stage, it is wise for the designer to hold preliminary meetings with the local authority to assess the likelihood of approval, to define any agreements the employer may have to enter into with the authority and to determine the timescale involved. If reaction is unfavourable or objections from others seem likely, the designer owes a duty to the employer to advise him accordingly, indicating the possible extent of negotiation that may be required to achieve a consent, or the implications for time, cost and risk of taking a project to appeal.

In submitting a detailed planning application the design team are likely to have to produce the following information:

- Floor layout plans;
- Typical sections showing proposed heights;
- All elevations;
- Elevations or pictorial views showing the proposals in context with adjoining buildings, where relevant;
- Site plans showing relationship of the proposal to other buildings, orientation, access, parking standards and the like;
- Details of materials and colours.

This information should be legibly and accurately presented in view of the different groups, committees, departments and members of the public that may have the right or duty to give their opinion.

If the application is refused, or the application is not determined within an agreed timescale or the employer objects to certain conditions imposed on a consent, then the applicant has the right to appeal within a time limit. Great care should be taken in advising an employer in these matters, especially in view of the Government imposing greater penalties for unreasonable or unsuccessful appeals. The employer may be well advised to seek expert legal advice prior to pursuing an appeal if the proposal is complex or risky.

Once a detailed consent is granted an employer can instruct the project team to proceed to the production stage; if such instruction is given before consent the employer must accept the risk against possible abortive work.

Programming

At the time of detailed planning consent the design team can re-examine the outline programme made at the start of the project and amend it in the light of the stage reached, the procurement option chosen and any financing conditions.

The purpose of the detailed pre-contract programme is to set out a sensible and logical sequence of the various pre-contract operations appropriate to the members of the design team and the required level of their input, together with external factors particular to the project.

The initial barchart or network can be expanded to show a programme to include:

- Input by all the appointed consultants;
- Design team progress meetings;
- Building Regulations submissions;
- Negotiations with service authorities;
- Scheduling each drawing and specification required;
- Key dates for exchange of information;
- Commencement and completion of bills of quantities;
- Completion of specification packages;
- Return of specialist design elements;
- Return of sub-contract and suppliers' quotations;
- Obtaining legal agreements for access arrangements, party wall awards, statutory bodies and the like.

In conjunction with these key points, the team should build in a tolerance to cover any unknowns: time for co-ordination within the team and additional time for the inevitable drawing changes during the process and for checking completed documentation. For the programme to run smoothly, all team members including the individual staff involved should be fully aware of the programme in respect of their own role and the interdependence of members of the design team and the implication of one member's delay on the others. It should be the duty of the team to keep the employer informed of progress against the programme and to give advice on any factors which may alter the programme and their effect in cost or time.

The Project Leader

In Chapter 1 reference is made to the traditional role of the architect as project leader. However, increasingly with larger projects the employer may appoint a specific project manager to co-ordinate the pre-contract and site works on his behalf.

This leader should seek to manage the effective production of all the required information and ensure that the team is working cohesively towards the common goal. He should be aware, therefore, of the terms of appointment or limit of responsibility of each team member and ensure that they work together to produce a coherent design and to prevent any overdesigning.

It is easy for time to be wasted if the members of the team are not in regular contact. Meetings should be held to:

- Review any new information;
- Review the detailed design;
- Examine structural and service implications for the design;
- Instruct on any necessary changes to the setting out, layout or facilities;
- Discuss any specialist design input;
- Chart progress against the programme.

Clear decisions must be made at these meetings. Time spent constantly changing drawings is unproductive and can lead to abortive costs. A record should be kept of decisions made and action required from consultants, so that the line of responsibility can be maintained.

The design team members should not lose sight of the fact that their role is to provide a service to the employer. Professional practices are now adopting procedures and methods of working to ensure a good quality of service to the employer and to maintain good practice within the office. One such aid to good practice is the importance of allowing sufficient time for checking drawings, documents and schedules. There is often a need for an objective view to point out more practical construction details for example, or to identify simple errors or discrepancies between drawings, thereby reducing the risk of future problems.

Drawings

The most traditional and widespread method of conveying information for building is the drawing. Drawings are used to convey fundamental information. It is common for the architect to commence by producing basic line drawings of plans and sections (general arrangements) which can then be used for:

- Structural and service engineers to detail their designs;
- Setting out dimensions;
- Obtaining specialist details and prices;
- Building Regulations submissions;
- Revising cost plans.

Once the setting out, structure and service allowances have been agreed, the designers then proceed to the production of large-scale details and junctions.

The clarity of drawings or documents is fundamental to the smooth running of a project. The project information should be presented to allow for easy understanding and assimilation both by fellow team members and by the contractor's site personnel, using workable cross-referencing systems where necessary for co-ordinated project information.

Specifications

An important role of the design team, and the architect in particular, is as specifier. In the early stages of the pre-contract programme, the designer will need to discuss and agree the types of materials and quality of installations and finishes.

This responsibility will rest with the specifier to determine in precise description the materials, fittings or installations which:

- Are appropriate to function, exposure or predicted use;
- Are available without undue complications;
- Meet levels of heat loss and sound attenuation;
- Have acceptable behaviour in fire;
- Are environmentally friendly in manufacture and use;
- Create the required ambience (colour, texture, space);
- Are energy-efficient;
- Are easy to maintain or clean;
- Are appropriate to the employer's budget.

The details and quality of the materials and construction required by the designer, translated into written form, comprise the specification. Methods of writing specifications, the different types of specification and their use as contract documents are discussed fully in Chapter 5.

Bills of Quantities

The decisions made by the specifier in conjunction with the design team and importantly the employer, form the basis of the written contract documentation. The method for presenting this information should be decided at the start of the programme and will depend upon the type of procurement route.

The use of bills of quantities can give rise to a better financial breakdown of the tender price by requiring the contractors to cost individual measured items of work. The production of the bills requires the quantity surveyor to translate the drawings into separate areas of work or sequences of work. Chapter 6 explains in detail the different types of bill and their appropriate function.

Specialist Sub-Contractors and Suppliers

In a project which demands specific fittings or materials, or special types of installation not considered within the main contractor's scope, the design team would normally approach suitable sub-contractors or suppliers for relevant design input and quotations. The method for including this information in the contract documentation is explained in Chapter 7. Once again, the onus is on the co-ordinator for this specialist information to be obtained within the programme.

Quality Assurance

Throughout all procedures the quality of the product or service should be at an acceptable level. Quality assurance is a management process designed to provide a high probability that the defined objective of the product or service will be achieved.

To achieve quality consistently, an organization needs to have a management system in place. That system should be capable of applying appropriate management checks throughout all work stages, be it a design service, a product manufacturing process or a building erection process, and of correcting deficiencies if they occur.

Ground rules designed for manufacturing industry are laid down in BS 5750:1987 which by careful interpretation can be adjusted to suit professional services. The professional institutions have studied the British Standard, which mirrors ISO 9001, its international counterpart, and have published recommendations to their members on quality assurance.

A recognized way for a practice to set up its own quality system is to prepare a quality manual covering the following:

- Overall policy of the practice as regards quality of service.
- Policies on such matters as information services, staff training, resource control, and documentation.
- The preferred methods of running projects, such as those based on the RIBA Plan of Work or some other procedures manual.

The principals of the practice should formulate, endorse and evaluate the manual and communicate its contents to all personnel, ensuring full understanding.

A practice may then decide whether or not they wish to seek third party assessment by a certification body leading to registration under BS 5750. This body will first examine whether the manual prepared by a firm meets all the requirements of the British Standard and then whether the firm is operating the manual correctly and thoroughly. After registration, further third-party assessments are usually carried out every four months to ensure that the firm is consistent in its operation of the manual. While certain employers, both public and private, are already requesting statements of QA from their consultants, the value of quality assurance registration to a professional practice may not be on a par with its use to product manufacturers.

Quality assurance on its own will not improve a firm's professional standards: it only provides an auditable record of performance against the firm's stated objectives, which may or may not embody high professional standards. The adoption of QA will promote consistency in performance at the level set down within the quality manual.

Obtaining Tenders

This chapter has discussed the various elements involved in the pre-contract programme and factors important to the smooth running of that programme. It has also examined the different types of information that are used to convey to the contractor the work to be done and to obtain tender prices. These different types of information are now discussed in greater detail in the following chapters with the final one recommending the procedure for obtaining selective tenders. The Aqua book *Contract Administration* then takes up the reins and guides the design team through the roles on site until the project is complete, while the Aqua book *Tenders and Contracts for Building* provides alternative methods of obtaining tenders.

Chapter 3

Approximate Estimates, Cost Planning and Control

Costs

One of the most important aspects of a proposed building project will be cost. The employer will want to know what he has to pay and when. In this chapter the word 'cost' applies to the cost of the work for the employer. 'Price' would be a more accurate term, but traditionally this information is referred to as approximate estimate or cost, cost planning, cost control and cost use. Cash flow is a true reflection of its meaning.

Each cost operation involves different and separate considerations. There are situations where cost is used in determining the financial feasibility of purchasing land or buildings in property development or redevelopment. In these cases the approximate cost is often required urgently. The scope of detail available may be limited and this underlines the essential need to establish in clear terms the basis on which costs are reported and the degree of approximation resulting.

Design a building regardless of cost

Basis of Reporting

It is unlikely that all the essential items are available as the basis of reporting costs. In any event, each project will have its own particular restrictions or requirements. The following list is not comprehensive, but it does indicate the kind of information to be determined and recorded in all types of pre-contract cost reporting.

- Name of employer.
- Address and extent of site and restrictions thereto.
- Names of design team members.
- Date of report and reference to the date on which the cost is based and indications of cost increases during periods of rapid inflation.
- Date and period of proposed building contract.
- Method of procurement of contractor.
- Funding of finances.
- Design and construction criteria including scheduling drawing numbers and reference to outline specifications or notes.
- Details of phasing of the construction.
- Availability of all supply services (water, gas, electricity, drainage).
- The planning situation, whether approval in outline or detail has been given and the influence on costs of planning requirements such as section 52 of the Town and Country Planning Act.
- The situations relating to preliminary work such as site survey, ground investigation, pre-contract demolitions and ground improvements.
- The need for early orders for items requiring long delivery periods.
- The inclusion or exclusion of value added tax, professional fees, building inspection and planning fees and similar items.
- The inclusion or exclusion of fitting out the building and installation of special requirements such as computer and data lines, sprinklers, dry risers, fire-fighting equipment, public address and telephones.
- A programme showing when full design instructions to the contractor should be achieved to prevent additional construction costs from disruption through lack of providing information at the proper time.

Good communication is fundamental and cost reports should be clear, unambiguous and leave no doubt in the reader's mind as to the basis of the information. The basic operations involved in providing cost control and advice are:

- Approximate estimate;
- Cost plan;
- Cost control;
- Cost in use;

- Estimated cash flow.

Before considering these operations in detail the integration of cost into the design process is illustrated in the diagram opposite.

The procedures to be followed in approximate estimating and cost planning will be governed to a large extent by the employer's attitude to the financial aspects of the proposed development. In most cases the approach to the finances of a project comes under one of three headings, as follows:

(1) *Cost limit or fixed capital expenditure.* In such a case the employer will state in his initial brief to the architect how much he is prepared to spend and will require the most suitable building he can obtain for that sum.

(2) *Unit cost limits.* These apply usually in the case of public bodies or other employers responsible for much repetitive building work, standard cost limits being applied on a unit basis, such as £x per person accommodated or £x per square metre of floor area. The best building must then be provided within these limits.

(3) *Budget, based on estimated cost of proposed development.* This is probably the most common approach in the case of the private developer. In the first instance he obtains guidance on costs from the architect and quantity surveyor from which he can make his first decisions regarding the brief to the architect and establish the cost limit for the job. This may involve him in considering such matters as the availability of the capital required, the likely return on the capital as an investment, or the economics of the proposed development in relation to his business.

From the foregoing it can be seen how important is the accuracy of the approximate estimate given in the early stages and also the adequacy of cost planning and control during the pre-contract period.

Approximate Estimate

Approximate estimates provide the architect with the anticipated costs of the whole or part of the project at various stages before tenders are received. Historically and contractually, the architect is aware of and controls the cost of building work through design and construction detailing. As cost information is provided, this design and detailing can be modified to suit the employer's requirements. There is no rigid classification of the types of approximate estimate but the following are in general practice; to a certain extent the four types of estimate described overlap one with the other.

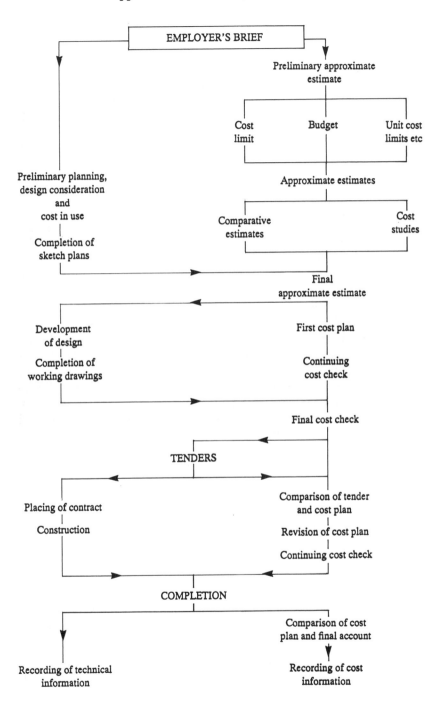

(1)   *Preliminary approximate estimate.* This estimate may be derived from a rough floor area, cost per head or other similar unit method based on previous experience, taking into account site conditions and general cost trends in the industry. It will be the earliest estimate given and is only an indication to ensure that the project is of the same financial order as that contemplated by the employer. It should be remembered, however, that of all the cost information given to an employer during the planning of a building, the figures given in the first approximate estimate are the ones that remain most clearly in his mind. Furthermore, it is frequently upon these figures that the whole economics of a development are based.

In this case, the estimate may have to be given without any drawings; nevertheless whether simple sketch plans are available or not the essential information required is as follows:

● Type of building and its use;
● The total floor area, overall height and number of floors;
● The site and its nature;
● An indication of the quality of work to be specified;
● A brief outline of the engineering services.

(2)   *Approximate estimate based on floor area with appropriate parts taken from approximate quantities.* This will be more accurate than (1) and may be used as confirmation or otherwise of a previous estimate prepared when less information was available. If information is taken from a similar job it may be as accurate an estimate as is required before tenders. In this form of estimate only the building should be based on floor area. External works should be based on approximate quantities and any special or unusual items should be measured out. Specialist services should also be dealt with separately. This approximate estimate may be used for establishing a total cost which is not to be exceeded.

The information to be provided here would be:

● Sketch plans:
    1:200 or 1:100 plan of each floor;
    1:200 or 1:100 elevations of most faces;
    1:200 or 1:100 sections;
    1:500 or 1:200 site plan showing extent of external works.
● Specification notes:
    Type of structure;
    Materials for walls, floors and roof;
    Floor loading;
    Column spacing (if not shown on drawing);
    General standard of finishings to walls, floors and roof;
    Types of windows and doors;

Type of stairs;
Type of foundations;
Extent of joinery fittings required;
Type and scope of engineering and specialist services including:
heating and hot water installation; electrical installation; mechanical
ventilation; sprinkler system; lifts and escalators;
other services such as gas, vacuum, compressed air.
- External services including positions of connections to all service mains and sewers.
- External works including types of road or paving, boundary walls, fences, gates.
- Any unusual conditions of work or contract.

(3) *Approximate estimate based on approximate quantities.* With certain types of work, especially alteration work or unusual types of construction, or when relevant information from a similar job is not available, an approximate estimate should be prepared on the basis of approximate quantities with each item being priced. Approximate quantities can be inclusive of all labour items, and where detailed construction represents relatively small expenditure, then cost items covering all work would be included. In many buildings, the major cost factors are represented by probably as few as 50 to 100 items. If these are accurately measured and priced this is likely to represent some 90% of the total cost leaving the remainder to be assessed in a more general way.

The information required would be similar to that listed in (2) but with the addition of:

- Sketch plans or preliminary working drawings:
1:100 plans of each floor;
1:100 elevations of each face;
1:100 sections;
1:500 or 1:200 site plan showing extent of external works;
1:20 typical sections through building;
Typical details of important features.

(4) *Approximate estimate from pricing an accurate bill of quantities.* This will give the most accurate approximate estimate possible but will normally only be available just before tenders are received. It is valuable as a check on the lowest tender and particularly useful in proving the details of the cost plan or giving advanced warning where adjustments in the construction may be necessary. The information required for this exercise are the bills of quantities and all information provided to the contractors tendering.

After completing the preliminary approximate estimate, various general options may be made the subject of cost studies and then comparative approximate estimates may be prepared. These might deal with such matters as alternative shapes of plan, number of storeys, types of structural frame and other matters fundamental to the design.

When all concerned are satisfied with the general manner in which the employer's requirements can be met, sketch plans and approximate estimates can be finalized and, as a prelude to the preparation of working drawings, the cost plan can be prepared. The latter may well have been started in conjunction with the cost studies previously referred to.

Cost Plan

The cost plan is the instrument or document which brings design and cost together at pre-tender stage. Each project comprises a series of elements which in themselves have particular design or specification characteristics. Although elements can influence each other, each will have a high percentage of its cost determined by the selection of material or design factors related directly to that element. At first, therefore, the element can be considered separately so that design, specification and costs become acceptable; influences on other elements will need to be considered afterwards. Using the latest approximate estimate of costs, each element can be separately valued.

In order to prepare a cost plan it is first necessary to sub-divide the building into a series of elements. Choice of elements will be based on the nature of the building and on any particular services or features it contains. It is essential, however, to use the same basic elements wherever possible in order to maintain continuity in compiling cost information.

The accuracy of the first cost plan will depend on when it is prepared. If it is compiled at a very early stage, then considerable variation could occur as the detailed design of the project proceeds.

Any number of alternative designs and specifications may be considered for each element. These can be valued for the purpose of comparison and to enable the architect to see how various design decisions may affect the cost and how he can make the best use of the employer's money.

By the time the tenders are due to be received, the cost plan should have been revised so that financially it corresponds with the expected lowest tender figure. This will highlight in advance the elements which may need re-consideration should the tenders prove too high. This will greatly assist in reporting to the employer where variations could be made to adjust the tender submissions. As soon as a tender is accepted, whether the submitted tender is amended or not, the cost plan should be revised to coincide with the contract sum.

As previously stated, the choice of elements will depend on the nature of the building work. The RICS publish their Standard Form of Cost

Analysis and this provides a reasonably comprehensive check list of elements and sub-division. In some respects it does not reflect modern design; however, summary of this publication is set out below for reference and as an invaluable *aide-mémoire*.

---

1. SUBSTRUCTURE

All work below underside of screed or where no screed exists to underside of lowest floor finish including damp-proof membrane, together with relevant excavations and foundations.

2. SUPERSTRUCTURE

2.A. Frame

Loadbearing framework of concrete, steel or timber. Main floor and roof beams, ties and roof trusses of framed buildings. Casing to stanchions and beams for structural or protective pruposes.

2.B. Upper Floors

Upper floors, continuous access floors, balconies and structural screeds (access and private balconies each stated separately), suspended floors over or in basements stated separately.

2.C. Roof

2.C.1. Roof structure

Construction, including eaves and verges, plates and ceiling joists, gable ends, internal walls and chimneys above plate level, parapet walls and balustrades.

2.C.2. Roof coverings

Roof screeds and finishings. Battening felt, slating, tiling and the like. Flashings and trims. Insulation. Eaves and verge treatment.

2.C.3. Roof drainage

Gutters where not integral with roof structure, rainwater heads and roof outlets. (Rainwater downpipes to be included in 'Internal drainage' (5.C.1.).)

2.C.4. Roof lights

Roof lights, opening gear, frame, kerbs and glazing.
Pavement lights.

2.D. Stairs

2.D.1. Stair structure

Construction of ramps, stairs and landings other than at floor levels.
Ladders.
Escape staircases.

2.D.2. Stair finishes

Finishes to treads, risers, landings (other than at floor levels), ramp surfaces, strings and soffits.

2.D.3. Stair balustrades and handrails

Balustrades and handrails to stairs, landings and stairwells.

2.E. External Walls

External enclosing walls including that to basements but excluding items included with 'Roof structure' (2.C.1.).

Chimneys forming part of external walls up to plate level.

Curtain walling, sheeting rails and cladding.

Vertical tanking.

Insulation.

Applied external finishes.

2.F. Windows and External Doors

2.F.1. Windows

Sashes, frames, linings and trim.

Ironmongery and glazing.

Shop fronts.

Lintels, cills, cavity damp-proof courses and work to reveals of openings.

2.F.2. External doors

Doors, fanlights and sidelights.

Frames, linings and trims.

Ironmongery and glazing.

Lintels, thresholds, cavity damp-proof courses and work to reveals of openings.

2.G. Internal Walls and Partitions

Internal walls, partitions and insulation.

Chimneys forming part of internal walls up to plate level.

Screens, borrowed lights and glazing.

Moveable space-dividing partitions.

Internal balustrades excluding items included with 'Stair balustrades and handrails' (2.D.3.).

2.H. Internal Doors

Doors, fanlights and sidelights.

Sliding and folding doors.

Hatches.

Frames, linings and trims.

Ironmongery and glazing.

Lintels, thresholds and work to reveals of openings.

3. INTERNAL FINISHES

3.A. Wall Finishes

Preparatory work and finishes to surfaces of walls internally.

Picture, dado and similar rails.

3.B. Floor Finishes

Preparatory work, screed, skirtings and finishes to floor surfaces excluding items included with 'Stair finishes' (2.D.2.) and structural screeds included with 'Upper floors' (2.B.).

3.C. Ceiling Finishes

3.C.1. Finishes to ceilings

Preparatory work and finishes to surface of soffits, excluding items included with 'Stair finishes' (2.D.2.) but including sides and soffits of beams not forming part of a wall surface.

Cornices, coves.

3.C.2. Suspended ceilings
Construction and finishes of suspended ceilings.

4. FITTINGS AND FURNISHINGS
4.A. Fittings and Furnishings
4.A.1. Fittings, fixtures and furniture
Fixed and loose fittings and furniture including shelving, cupboards, wardrobes, benches, seating, counters and the like. Blinds, blind boxes, curtain tracks and pelmets. Blackboards, pin-up boards, notice boards, signs, lettering, mirrors and the like. Ironmongery.

4.A.2. Soft furnishings
Curtains, loose carpets or similar soft furnishing materials.

4.A.3. Works of art
Works of art if not included in a finishes element or elsewhere.

4.A.4. Equipment
Non-mechanical and non-electrical equipment related to the function or need of the building (e.g. gymnasia equipment).

5. SERVICES
5.A. Sanitary Appliances
Baths, basins, sinks, etc.
WCs, slop sinks, urinals and the like.
Toilet-roll holders, towel rails, etc.
Traps, waste fittings, overflows and taps as appropriate.

5.B. Services Equipment
Kitchen, laundry, hospital and dental equipment, and other specialist mechanical and electrical equipment related to the function of the building.

5.C. Disposal Installations
5.C.1. Internal drainage
Waste pipes to 'Sanitary appliances' (5.A.) and 'Services equipment' (5.B.).
Soil, anti-syphonage and ventilation pipes.
Rainwater downpipes.
Floor channels and gratings and drains in ground within buildings up to external face of external walls.

5.C.2. Refuse disposal
Refuse ducts, waste disposal (grinding) units, chutes and bins.
Local incinerators and flues thereto.
Paper shredders and incinerators.

5.D. Water Installations
5.D.1. Mains supply
Incoming water main from external face of external wall at point of entry into building including valves, water meters, rising main to (but excluding) storage tanks and main taps.
Insulation.

5.D.2. Cold water service

Storage tanks, pumps, pressure boosters, distribution pipework to sanitary appliances and to services equipment. Valves and tanks not included with 'Sanitary appliances' (5.A.) and/or 'Services equipment' (5.B.). Insulation.

5.D.3. Hot water service

Hot water and/or mixed water services.

Storage cylinders, pumps, calorifiers, instantaneous water heaters, distribution pipework to sanitary appliances and services equipment. Valves and taps not included with 'Sanitary appliances' (5.A.) and/or 'Services equipment' (5.B.). Insulation.

5.D.4. Steam and condensate

Steam distribution and condensate return pipework to and from services equipment within the building including all valves, fittings, etc. Insulation.

5.E. Heat Source

Boilers, mounting, firing equipment, pressurizing equipment instrumentation and control. ID and FD fans, gantries, flues and chimneys, fuel conveyors and calorifiers. Cold and treated water supplies and tanks, fuel oil and/or gas supplies, storage tanks, etc., pipework (water or steam mains), pumps, valves and other equipment. Insulation.

5.F. Space Heating and Air Treatment

5.F.1. Water and/or steam

Heat emission units (radiators, pipe coils, etc.), valves and fittings, instrumentation and control and distribution pipework from 'Heat source' (5.E.).

5.F.2. Ducted warm air

Ductwork, grilles, fans, filters, etc.
Instrumentation and control.

5.F.3. Electricity

Cable heating systems, off-peak heating systems, including storage radiators.

5.F.4. Local heating

Fireplaces (except flues), radiant heaters, small electrical or gas appliances, etc.

5.F.5. Other heating systems

5.F.6. Heating with ventilation (air treated locally)

Distribution pipework, ducting, grilles, heat emission units including heating calorifiers except those which are part of 'Heat source' (5.E.) instrumentation and control.

5.F.7. Heating with ventilation (air treated centrally)

All work as detailed under (5.F.6.) for system where air treated centrally.

5.F.8. Heating with cooling (air treated locally)

All work as detailed under (5.F.6.) including chilled water systems and/or cold or treated water feeds. The whole of the costs of the cooling plant and distribution pipework to local cooling units shall be shown separately.

5.F.9. Heating with cooling (air treated centrally)

All work detailed under (5.F.8.) for system where air treated centrally.

5.G. Ventilating System

Mechanical ventilating system not incorporating heating or cooling installations including dust and fume extraction and fresh air injection, unit extract fans, rotating ventilators and instrumentation and controls.

5.H. Electrical Installations

5.H.1. Electric source and mains

All work from external face of building up to and including local distribution boards including main switchgear, main and sub-main cables, control gear, power factor correction equipment, stand-by equipment, earthing, etc.

5.H.2. Electric power supplies

All wiring, cables, conduits, switches, etc., from local distribution boards, etc., to and including outlet points for the following:

General-purpose socket outlets.

Services equipment.

Disposal installations.

Water installations.

Heat source.

Space heating and air treatment.

Gas installation.

Lift and conveyor installations.

Protective installations.

Communication installations.

Special installations.

5.H.3. Electric lighting

All wiring, cables, conduits, switches, etc., from local distribution boards and fittings to and including outlet points.

5.H.4. Electric lighting fittings

Lighting fittings including fixing.

Where lighting fittings supplied direct by client, this should be stated.

5.I. Gas Installations

Town and natural gas services from meter or from point of entry where there is no individual meter: distribution pipework to appliances and equipment.

5.J. Lift and Conveyor Installations

5.J.1. Lifts and hoists

The complete installation including gantries, trolleys, blocks, hooks and ropes, downshop leads, pendant controls and electrical work from and including isolator.

5.J.2. Escalators

As detailed under 5.J.1.

5.J.3. Conveyors

As detailed under 5.J.1.

5.K. Protective Installations
  5.K.1. Sprinkler installations
    The complete sprinkler installation and Co_2 extinquishing system including tanks, control mechanism, etc.
  5.K.2. Fire-fighting installations
    Hosereels, hand extinquishers, fire blankets, water and sand buckets, foam inlets, dry risers (and wet risers where only serving fire-fighting equipment).
  5.K.3. Lightning protection
    The complete lightning protection installation from finials and conductor tapes, to and including earthing.
5.L. Communication Installations
  The following installations shall be included:
  Warning installation (fire and theft)
  Burglar and security alarms.
  Fire alarms.
  Visual and audio installations

Door signals.	Public address.
Timed signals.	Radio.
Call signals.	Television.
Clocks	Pneumatic message system.
Telephones.	

5.M. Special Installations
  All other mechanical and/or electrical installations which have not been included elsewhere, e.g. chemical gases: medical gases: vacuum cleaning: window cleaning equipment and cradles: compressed air: treated water: refrigerated stores.
5.N. Builder's Work in Connection with Services
  Builder's work in connection with mechanical and electrical services.
5.O. Builder's Profit and Attendance on Services
  Builder's profit and attendance in connection with mechanical and electrical services.

6. EXTERNAL WORKS
  6.A. Site Works
    6.A.1. Site preparation
      Clearance and demolitions.
      Preparatory earth works to form new contours.
    6.A.2. Surface treatment
      The cost of the following items shall be stated separately if possible.

Roads and associated footways.	Games courts.
Vehicle parks.	Retaining walls.
Paths and paved areas.	Land drainage.
Playing fields.	Landscape work.
Playgrounds.	

    6.A.3. Site enclosure and division
      Gates and entrance.
      Fencing, walling and hedges.

6.A.4. Fittings and furniture
Notice boards, flag poles, seats, signs.
6.B. Drainage
Surface water drainage.
Foul drainage.
Sewerage treatment.
6.C. External Services
6.C.1. Water mains
Main from existing supply up to external face of building.
6.C.2. Fire mains
Main from existing supply up to external face of building; fire hydrants.
6.C.3. Heating mains
Main from existing supply or heat source up to external face of building.
6.C.4. Gas mains
Main from existing supply up to external face of building.
6.C.5 Electric mains
Main from existing supply up to external face of building.
6.C.6. Site lighting
Distribution, fittings and equipment.
6.C.7. Other mains and services
Mains relating to other service installations.
6.C.8. Builder's work in connection with external services
Builder's work in connection with external mechanical and electrical services, e.g. pits, trenches, ducts, etc.
6.C.9. Builder's profit and attendance on external mechanical and electrical services
6.D. Minor Building Work
6.D.1. Ancilliary buildings
Separate minor buildings such as sub-stations, bicycle stores, horticultural buildings and the like, inclusive of local engineering services.
6.D.2. Alterations to existing buildings
Alterations and minor additions, shoring, repair and maintenance to existing buildings.

7. PRELIMINARIES

Cost Control

As the design is developed and decisions are made, so the cost plan must be checked to ensure that such decisions will not adversely affect the intended expenditure. Where the value of any element in the cost plan is seriously altered by a decision taken at this stage it will be necessary to review the value of other elements in the cost plan.

Close liaison between the architect and other members of the design team during this process will enable all concerned to be kept informed of any matters which might affect the cost of the building and to take such steps as may be necessary to ensure that the authorized expenditure is not exceeded.

In this book, events after the receipt of tenders are not dealt with, but it should be borne in mind that the process of cost control should continue throughout the period of the contract when variations should be considered in the same way that design decisions are considered during the pre-contract period. On completion of the contract a similar comparison should be made of the cost plan and the final account so that all matters relating to the cost of the building can be properly recorded to provide detailed cost information for use on subsequent occasions.

The processes of cost control are continuous so that the estimated final cost is always known. Quantity surveyors will document this information on standard forms or more likely through computers. No machine, however, will ever be able to replace the skill of an experienced professional and the implications of cost control should always be properly resourced and presented in a manner that is totally understood by the whole of the design team.

Cost in Use

Throughout the design period, reference will be made to cost in use. This can apply to total design or to the specific use of alternative materials. Total design being affected by cost in use can be illustrated by the following:

- Single or multi-storey design; low- and high-rise; open or cellular plan and the effect of these items on the effective use of the building.
- Fire insurance rating of the building; use of sprinklers and lightning conductors; restriction of floor areas or columns in one fire compartment; 'standard' of building to comply with the Loss Prevention Council Rules.
- Cost of fuel to heat or ventilate the building; provision of insulation beyond building control requirements; building management systems; choice of wall and roof cladding materials.
- Ground improvements; dynamic or vibro compaction; piling; floor loadings; being taken into account in the value or purchase price of the land.
- Financial return on investment; planned useful economic life of the building to the present or future occupier; air conditioning; raised access floors; provision of crèches, swimming pools and other social amenities.

As each design element is considered, cost in use will apply to the choice of materials. Almost every situation will produce alternative selections each having a differential cost to be taken into account when assessing its use or life. Examples can vary from major items such as the cost of main sewer connection compared with cesspool drainage and cost of frequent emptying, to lesser items such as toilet wall plaster and painting, with frequent redecoration costs, compared with wall tiling.

A flexible judgement should be made on cost in use, particularly on major items. Traditional construction such as private housing tends not to change over a decade or two, but other types often do. Industrial, office and speculative buildings have relatively short lives and become out of date or redundant. Cost in use should properly anticipate the effective life of the building and not allow design to incorporate expensive selections which may not have time to give the financial return.

Cash Flow

Cash flow is a forecast to show when sums of money are required to be provided by the employer. The capital cost of a building project will be funded from reserves of the employer or loans from banks or institutions. When money is required to pay certificates and invoices the employer will need to know the dates in advance so that he can safeguard unnecessary loss or charge of interest.

Forecasts of cash flow to numerous parties could be required by the employer, the most likely being:

● Land where purchased by instalments;
● Certificates for contractor;
● Payment to statutory authorities where directly appointed by the employer;
● Invoices from consultants;
● Fees for planning and building control;
● Use of services (gas, electricity) during building work to existing premises;
● Interest charges on funding capital.

The most common cash flow request relates to payments to the contractor and this could be shown as follows:

Name of Project

Partial Fitting Out

Cash Flow Forecast for Contractor's Building Work

September 1992

Payment will fall due during the *first 10 days* of each month shown.

		£
1993		
January –	Contractor commences site work	—
February		—
March		130,000
April		240,000
May	(Practical completition)	280,000
June	(First moiety released)	220,000
July		27,000
1994		
May	(Second moiety released and final certificate)	23,000
Pre-contract cost plan		£920,000

The above is a very simple example of a cash flow at pre-contract stage to correspond with the cost plan also at pre-contract stage. When the work proceeds to contract the cash flow will be amended to total the contract sum and the dates adjusted to suit the dates of possession and practical completition stated in the contract. Should the contract sum vary considerably owing to variations or extension to the contract period, this will require the cash flow to be revised accordingly.

Chapter 4

Drawings and Schedules

Quality

Drawings represent the most important means by which the designer conveys his intention to the employer, statutory authorities, consultants, quantity surveyor, contractor and sub-contractors and, as has been stated in Chapter 2, the absolute clarity of drawings is essential. This cannot be overemphasized. Clear, concise, well-planned, co-ordinated drawings not only make the information they contain easy to understand, they also inspire confidence. Conversely, poor drawings do little except reveal the designer's lack of knowledge and inability to conduct affairs in a well-ordered manner.

Recommendations on the preparation of drawings, especially working drawings, are contained in BS 1192:1984. Also important are the principles defined in BS 5750:1987 on quality assurance concerning the orderly preparation, dissemination and recording of design data and indeed all documentation relating to any form of manufacturing and production.

The method of producing drawings will vary from office to office depending on the size and resources of the practice. Some may rely heavily upon the use of computer-aided design, others on traditional draughting techniques, but most will rely on a combination of the two. The type of employer, the nature of the project, the degree of involvement with other consultants and their resources and, finally, the programme for the work, all determine how the drawings at each stage are to be dealt with. But if quality is to be assured, there are simple rules that need to be observed no matter how the drawings are produced.

Types, Sizes and Layout of Drawings

The types of drawings vary as the project proceeds through its programme, as is described later in this chapter. However, for efficiency and economy in the use of time, all drawings should be:

- On standard-sized sheets laid out and pre-printed in such a manner that the source and purpose of the drawing can be readily identified;
- Shown to contain all necessary routine information and able to be readily checked;
- Kept in comprehensive sets and stored easily.

An assortment of drawings of different shapes and sizes with title panels and essential information in differing positions is a cause of confusion and irritation. Standardization should apply no matter how the drawings are prepared.

In BS 3429:1984 recommendations are made for drawing sheet sizes A0, A1, A2, A3 and A4. In selecting the standard drawing sizes, account should be taken of the many and varied methods of reproduction: dyeline printing, photocopying, laser printing and photographic reproduction. Excluding routine dyeline printing, all methods available allow the possibility of changing the scale of the drawing in the course of reproduction: an invaluable asset in planning and producing comprehensive and co-ordinated sets of drawings. With this advantage in mind it is probable that the range of standard sheets might reasonably be limited to A0, A1 and A3.

A0 is a rather unwieldy size, especially for handling on site, but may be necessary for general arrangement drawings of large projects, extensive landscape drawings and the like. A1 size drawings are the most common and the most acceptable for ease of handling and can be reduced to any size down to A4. This is extremely useful in the preparation of design and presentation drawings that may be required in bound folders of A3 or A4 size. A3 sheets are most commonly used for quick sketches and details that may be reduced down to A4 size for use in transmission by fax.

With the almost universal use in offices of personal or mainframe computers, the use of drawing sheets for the preparation of schedules should be the exception rather than the rule. It is likely that standard letters, forms of instructions, certificates and most schedules would be held on disc for completion or amendment as the occasion arises. Drawn schedules might be used only for window or door and ironmongery schedules which contain a considerable amount of information; in this case A3 sheets would be the most convenient.

The design of drawings must take account of both immediate and long-term storage. Whether it is for reduction and storage in files, in plan chests or in large clips hung on racks beside working stations, the layout of the sheet must provide sufficient margin to ensure that no information is obscured by the storage system.

Two types of title and information panel are recommended in BS 1192. The examples include space for CI/SfB* references, but this will not be required unless a CI/SfB classified form of specification is being

---

*CI/SfB is a co-ordination system developed for the construction industry. It allows cross-referencing between drawings, specifications and technical literature. The CI/SfB reference is most commonly seen in the top right-hand corner of trade literature. It originated in Sweden, and in 1968 the RIBA developed the system further and added the prefix 'CI' (standing for 'construction indexing') to distinguish this system from the original SfB.

The recording of information . . . presupposes that someone will want to refer to it later

employed. An alternative classified form is CPI (co-ordinated project information), a comprehensive coded system provided an interface between the manufacturer's product database, the use of materials and methods of assembly shown on the drawings, the specification and the bills of quantities, using SMM7 (see also Chapter 6 under the heading Additional Information).

Seldom are drawings produced and left unchanged. The development of the design, collaboration between the architect, structural engineers, service engineers and specialist sub-contractors and suppliers, all lead to a continuing process of amendment and revision. For the safety of co-ordination every revision to a drawing must be noted on the sheet and indicated by a revision number. Unless amendments are adequately described and the date when they were made noted, it can be difficult for other members of the design team or the contractor to discover changes made to the drawings.

Drawings are distorted by the printing process. It is therefore unwise to take dimensions off drawings by scaling. Also, drawings may be reduced on to smaller sheets for convenience or for the purposes of transmission by fax. This is often done without regard to the effect upon its scale. Also, drawings might be microfilmed for long-term storage of 'as built' information. In the face of all these possibilities, provision must be made for clearly recording not only the original, but also the reduced scale, and it must be indicated if a drawing is no longer to scale should it have been arbitrarily reduced in size. A drawn scale on a standard sheet can therefore be an invaluable reminder and tool.

Scale

Difficulties arise if drawings are prepared to unusual scales. It is important to use scales which are in common use. The following are recommended:

- Locations plans   1:2500
                                     1:1250
                                     1:500
- Site and development plans   1:500
                                     1:200
- General arrangement drawings 1:100
                                     1:50
- Component drawings   1:50
                                     1:20
- Details   1:5
                                     1:1 (full size)
- Assembly drawings   1:20
                                     1:10

Engineering consultants frequently use the scales of 1:25 and 1:250. This can cause considerable confusion and should be discouraged by the architect. Another confusing scale is 'half size' as it is too easily mistaken for full size.

Programme and Priority of Drawings

The type of contract and the method of choosing the contractor can affect the sequence of events and programme for the production of drawings. The building consists of many elements: structural frame, walls, partitions, roof, heating, lighting, ventilation and plant, plumbing and so on. Each of these elements forms a part of the whole and only a complete set of drawings can inform the contractor on the complete building. That does not necessarily mean to say that all drawings need to be completed before the contractor starts work on site: hence the need to understand at the outset the type of contract to be used.

The production of the drawings should be carefully planned. A preliminary list should be prepared of the drawings that will be required, with a programme that includes not only the preparation of the architect's drawings but also those of the other consultants. This programme will be determined by two basic forms of approach to procurement: (1) drawings for tendering and construction; and (2) drawings prepared for design and build or management contracts.

Drawings for Tendering and Construction

Where tenders are to be obtained and a contract placed on the basis of a JCT Standard Form of Contract, with or without bills of quantities, it is presupposed that all the drawings, schedules and specifications will be completed prior to obtaining tenders. It is useful to refer to the Outline Plan of Work (see Chapter 1). This will show how important it is to plan the production of the drawings, particularly through work stages E and F. It should also be borne in mind that where bills of quantities are used the SMM has general rules setting out details of the drawings required for the purposes of tendering.

The drawing sequence is as follows:

(1) *General arrangement drawings.* Plans and elevations (1:100 or 1:50) which, after receiving the employer's approval, are sent to all consultants who will then prepare their draft schemes.
(2) *Services drawings by the architect (plumbing, drainage).* To be worked out in detail concurrently with the preparation of the general arrangement drawings.
(3) *Construction details by the architect and specialist sub-contractors.* The selection of materials and finishes made during the preparation of the general arrangement drawings leads to the preparation of the architect's details and to obtaining tenders for specialist sub-contract design and construction. The outcome of this stage of the work can affect certain aspects, particularly critical dimensions in the primary elements of the structure and consequently the work of other consultants.
(4) *Assembly details.* When the consultants' drawings are accepted, the assembly details (1:20 and larger if necessary), which have been drafted in outline, should be completed.
(5) *Final co-ordination of drawings.* Now that all the detailed information has been assembled and co-ordinated, the final overall picture (originally the outline general arrangement drawings) can be completed with accuracy.
(6) *Layout and site plans.* Finally completed, incorporating information on all external services and the setting out of the buildings.

It is important to keep a record set of the drawings sent out to tender.

Many details and drawings will be changed during the course of the contract and, in adjusting the costs, it will be necessary to refer to the original information upon which the tender and contract sum was based.

Drawings Prepared for Design and Build or Management Contracts

There are circumstances which require the contractor to be responsible for the production of the building design or to manage a contract in such a manner as will accelerate the rate of construction (even though it might increase the cost). The appropriate forms of contract for such a situation are described in the Aqua book *Tenders and Contracts for Building*. The preparation of the drawings will be geared, not to the process for obtaining tenders for the whole of the works to be carried out by a single contractor, but to the alternative process of obtaining sub-contractors' tenders for the various parts of the building, known as 'packages'. The tenders for these packages will be obtained in a sequence related to the programme of work to be carried out on site, on the basis of drawings and specifications for each particular package. The order in which the drawings are produced will be related to the order of procurement and construction.

Once the building is designed, the sub-contract packages involved will be identified. The general arrangement drawings will have the same significance as in the traditional procurement route but, once they are agreed, subsequent drawings will be of an elemental nature: that is, concerned with the various and separate elements of the building. Pile or foundation drawings, together with the main frame and drainage drawings, will be completely detailed and the work started on site possibly before detailed or assembly drawings have started for secondary components of the superstructure.

Needless to say, the co-ordination of details in these circumstances becomes very difficult and can result in a rather different approach to design, calling for greater flexibility in the use of the structure or the building envelope and more readily allowing substantial changes to the design late in its development. The following is an example of the order in which drawings and specifications for packages might be required, those items marked with an asterisk (*) having a long lead-in or delivery period. There can be many more types of package and some may be combined.

(1)   General arrangement drawings;
(2)   Levelling and general site works;
(3)   Structural frame (steelwork)*;
(4)   Drainage;

(5)   Foundations (piling);
(6)   Lifts and escalators*;
(7)   Proprietary walling*;
(8)   Roofing systems;
(9)   Windows and doors*;
(10) Brickwork details;
(11) Mechanical and electrical engineering services*;
(12) Plant and machinery details;
(13) Partitions;
(14) Staircases and other secondary elements;
(15) Ceilings;
(16) Fixtures and fittings;
(17) External works.

Computer-Aided Design

The title given to this technique describes well the role of computers in the design process. Computers aid but they do not perform design tasks. Indisputably, computers are invaluable tools with many attractions to all the members of a design team. The main advantages lie in their ability to perform the technical services of producing drawings at great speed, of calculating and co-ordinating and ultimately of communicating. If there is a disadvantage, it may be seen in the type of drawings most commonly produced: of single line weight without enhancement or emphasis and often, for all their precision, lacking the special quality that allows easy reading and interpretation.

The most simple computer system will produce two-dimensional drawings containing no more information than is input for each drawing. The most advanced will store a complete model of the building and, on command, produce drawings to any scale or projection, of any component part, section, or the whole of that building, while amendments to the original model will automatically appear on all drawings as they are required and produced. Moving three-dimensional presentations in line or colour have a fashionable, persuasive quality. Also, the storage capacity of even modestly powerful computers can carry a vast library of standard components, assemblies and details that will enable the architect or technician to quickly produce a vast range of working drawings of great accuracy.

Where members of the design team are using compatible hardware, discs can be exchanged giving each designer immediate access to all the information produced by the others. Layer upon layer, information can be added to the general arrangement drawings ensuring complete co-ordination of all the parts of the building: structure, envelope details, mechanical and electrical engineering services and finishing details.

At the outset of the design process, members of the design team should discuss and agree the management of their CAD resources. All members of the team should produce one or more back-up copies of all drawings at all stages of production. Records of drawings undergoing major amendment should be stored. Common systems for coding drawings might well be adopted. In large projects where many thousands of drawings may be produced, it is possible for the designer to have his own mainframe system or to be linked through to specialist agencies responsible for cataloguing and distributing to those who need the many drawings as they are produced or amended.

CAD enables every conceivable calculation related to the design process to be carried out, including structural calculations, thermal performance and energy management calculations, fire safety and lighting. And by following the procedures of CPI, the designers' programs can be interfaced with those of the quantity surveyor so that coded information on the drawings can generate the production of the bills of quantities, specification documents and ongoing cost estimates.

While it might seem but a small step to the creation of hardware and software that will actually carry out the design function in response to no more than verbal commands, the reality of CAD in practice is rather different.

Although a large number of practices are now operating and relying on CAD, the cost of equipment, the programs and the specialized operators remains high. The investment can only be justified if the equipment is in constant use. Not all buildings are of a scope, size or type suitable for detail design using CAD. Large buildings, repetitive on many floors, are particularly suitable. The drawings for small buildings of a very specialist nature and perhaps irregular shape are more economically done by hand. The reason lies in the time required for setting up the building in its program. The same problem can arise in setting up amendments to the original drawings. Time lost at this stage needs to be recouped in the time gained in the rapid production of the finished drawings.

The use of CAD does not change the ground rules for good practice, but the systematic approach that it requires does help to improve the clarity of the information produced and makes it more accessible. Whichever way the drawings are produced, the information they convey should always be the same.

Contents of Drawings

When planning and programming the production of drawings, account must be taken of the fact that every part of the building and its site has to be designed, detailed and drawn. Obvious as this might seem, it is unfortunately only too common to find elegant sets of drawings repetitiously showing the simple but assiduously avoiding the complicated. Should a part of a building be difficult to detail and to draw, then that is the very detail and drawing that the contractors will need most. In considering how all the necessary information is to be conveyed, there are good and simple rules to observe:

- Draw every part of the building.
- Do not repeat information unnecessarily.
- Sections are invaluable – indicate and code their location clearly.
- Remember that plans are horizontal sections; one plan for each floor level may not be enough.
- Number all rooms and spaces.
- Give reference numbers to all doors, windows, built-in fittings and the like; show these on all plans, sections and elevations.
- Make clear cross-reference to other drawings or to schedules so that detailed information can be traced.
- Be consistent in showing structural grids and levels on all plans, sections and elevations.
- Provide concise notes laid out clear of drawn information.
- Work out and indicate all essential dimensions but do not labour the obvious.

The following is a checklist of drawings and their contents:

- **Survey plan**
  (1) Existing site, boundaries and surrounds.
  (2) Positions of major features such as existing buildings, roads, streams, ponds, walls and gates.
  (3) Positions, girth, spread and type of trees; location and type of hedgerows.
  (4) Sufficient spot levels and contour lines related to a specific datum, to enable a section of the site to be drawn in any direction specifically required.
  (5) Position, invert and cover or surface levels of existing drains; location, direction and depth of service mains.
  (6) Rights of way, bridle paths and public footpaths.
  (7) Access to site for vehicles.
  (8) Ordnance reference if available.
  (9) North point.
  (10) A key plan showing the relationship of the various sheets, should the survey cover more than one standard sheet.

- **Site plan, layout and drainage**
  (1) Relevant information from the survey plan.
  (2) Building profile and grid dimensionally related to a datum point or line.

(3)   New roads and paths with widths and levels marked.
(4)   Floor levels clearly indicated using the same datum as for the existing levels on the survey plan.
(5)   Steps and ramps where they occur.
(6)   Trees and hedgerows removed or retained.
(7)   Street and other external lighting.
(8)   Soil and surface water drains complete with pipe sizes and connections to sewers, making clear distinction between soil and surface water drains and manholes (manhole sizes, levels and invert levels should be shown on a separate schedule).
(9)   Gas, water and electric mains with depths indicated where possible and showing positions of:
    (a) connections to existing mains.
    (b) supply company's meters (external) and details of meter housings if required.
    (c) points of termination within buildings.
(10) Banking and cutting and areas of disposing or spreading surplus soil.
(11) Details of fencing, existing and new.

● **General arrangement drawings for services**
Once the design is approved and working drawings commenced, copy negatives should be taken of the general arrangement drawings of all floors showing only door swings in addition to walls and partitions and without any dimensions or other information on them. These drawings then can be developed without confusion or loss of time to show:
(1)   Incoming mains and meter positions for all services.
(2)   Electrical layout, including power points, light points, switches, and all electrical equipment.
(3)   Heating layout showing boilers, calorifiers, radiators, or similar heating equipment.
(4)   Plumbing and internal drainage layout.
(5)   Gas layout.
(6)   Sprinkler system layout including pumps and storage tanks.
(7)   Fire detection and protection equipment.
(8)   Location of any special services.

● **General arrangement drawings – foundations**
(1)   The main building grid annotated and dimensioned.
(2)   Width and depth of all foundations for walls, piers and stanchions, with levels to underside.
(3)   Positions and levels of drains, gullies and manholes close to foundations.
(4)   Walls above foundations dotted with wall thicknesses dimensioned.
(5)   Positions of incoming service mains, service main ducts and trenches and their levels.

- **General arrangement drawings – plans of all floors**
  (1) Complete plans drawn at a constant level through all openings at all floors and mezzanine floors.
  (2) Dimensions as follows:
      (a) overall dimensions.
      (b) external dimensions, taking in all openings.
      (c) internal dimensions to show the positions and thicknesses of internal walls, partitions and all major features.
  (3) Doors, their direction of swing and reference numbers.
  (4) Windows and their reference numbers.
  (5) Names or numbers of all rooms and circulation spaces (see BS 1192).
  (6) Vertical and horizontal service ducts, holes in floor slabs, flues, and builder's work in connection with services.
  (7) Staircases, their direction of rise and stair treads numbered.
  (8) Hatching to indicate materials of which walls and partitions are constructed.
  (9) Rainwater pipes.
  (10) Profile of joinery fittings and their code or reference numbers.
  (11) Sanitary fittings.
  (12) Floor levels clearly marked.

- **General arrangement – roof plan**
  (1) Main construction features.
  (2) Levels clearly marked.
  (3) Types of covering.
  (4) Direction of falls.
  (5) Rainwater outlets, gutters and pipes.
  (6) Rooflights.
  (7) Tank rooms, trap doors, chimneys, ventilation pipes and other penetrations; builder's work in connection with services.
  (8) Parapets, copings and balustrades.
  (9) Duckboards, catwalks, escape stairs and ladders.
  (10) Lightning conductors and flag poles.

- **Elevations**
  (1) Elevations of all parts of the building showing new and old ground levels, profiles of foundations, damp-proof course levels and levels of ground and upper floor slabs.
  (2) External doors and windows with their reference numbers.
  (3) Air bricks and ventilators.
  (4) External materials, flashings.
  (5) Rainwater pipes and gutters.
  (6) Lightning conductors.

● **Sections**
(1)   Sections that will describe all parts of the building and especially the relationship between separate but linked buildings.
(2)   New and original ground levels showing cut and fill.
(3)   The main structural profile with grid lines.
(4)   Type and depths of foundations.
(5)   Floor and roof structures.
(6)   Windows, doors and roof lights with their reference numbers.
(7)   Lintels, arches and secondary structural members.
(8)   Damp-proof membranes and courses, flashings.
(9)   Eaves and valley gutters, parapets, rainwater heads and down pipes.
(10) All vertical dimensions.

● **Ceiling plans**
(1)   Mirrored ceiling plans at all floor levels.
(2)   Room names and numbers over which ceilings occur.
(3)   Type of ceiling, type of suspension, suspension pattern and finishes.
(4)   Height of ceiling above floor level.
(5)   Location of light fittings, their type and size.
(6)   Fire detection and safety installations, ceiling void compartmentation, smoke detectors, emergency lighting.
(7)   Mechanical ventilation registers.
(8)   Sprinkler layouts.

● **Construction details (scale 1:20 and 1:10)**
(1)   Detailed sections for external walls, foundations and roofs.
(2)   Plans, sections and elevations for staircases.
(3)   Lift wells and escalators.
(4)   Any room or part of the building the setting out of which is difficult, involves extensive fittings, fixtures, plumbing or special features or requires careful co-ordination such as:
(a) kitchens.
(b) bathrooms.
(c) lavatories.
(d) special-purpose rooms as in hospitals, laboratories.
(5)   Building in window and door details.
(6)   Part elevations and sections of the building's envelope containing special features such as:
(a) entrances.
(b) special brick details.
(c) balconies.
(d) stonework.

(7) Plant rooms.
(8) Builder's work in connection with services.
(9) Vertical and horizontal pipe ducts and their access.
(10) Fireplaces and flues.
(11) External details such as special paving, steps, kerbs, handrails, and external lighting.

● **Large-scale details (scale 1:10 and 1:15)**
These comprise enlargements of component parts of assemblies which have been shown to 1:20 scale but which require a larger scale to show the full details.
(1) Cills, heads and jambs of windows and doors.
(2) Special brickwork, string courses, arches (rubbed bricks) and special flashing details.
(3) Eaves, parapets, copings and special mouldings.
(4) Timber sections such as handrails, window sections, and joinery details.
(5) Jointing details of curtain walling and other specialized cladding.
(6) Staircases.
(7) Special joinery fixtures and fittings.
(8) Any special feature which cannot be described or shown clearly to a smaller scale.

Schedules

It is good practice to convey information on items such as windows, doors and ironmongery by means of schedules which will set out the architect's requirements for other members of the construction team – in particular the quantity surveyor and later the contractor – in a manner which has many advantages over drawings; for example:

● Checking for errors of omission or duplication is simplified.
● Accounting of similar items for the purpose of obtaining estimates or placing orders is simplified.
● The omission of information about an item is unlikely as the appropriate space in the schedule would otherwise remain blank.
● If the information is set down systematically, prolonged searching through a specification is avoided.

In setting down information in schedules, ensure clear layout, simple coding, the minimum of abbreviation and proper reference to the location in the building of the items concerned. Failure to recognize undue complexity in schedules leads to information that might have seemed abundantly clear to the author but, to the man on the site, is shrouded in obscurity.

Schedules should be either the same size as the standard drawing sheets selected for the job or, if smaller, bound together as a set in one of the accepted A sizes.

The design of and information contained in the title blocks should be consistent with those of the drawings. Schedules should also be numbered in a manner which is an extension of the numbering system used for the drawings. The first schedule should be a list of all the drawings and schedules related to the project with space to record the latest revision number of each document. The first formal use of this schedule will be its inclusion in the bills of quantities providing a record of the drawings and schedules used in the preparation of the tender documents.

The layout of schedules will vary according to the information to be conveyed. Items being scheduled may be listed to read downwards on the left-hand side with the characteristics and sizes of the items reading across or vice versa. Diagrams may well be included in the body of the schedule and a column reserved for notes and records of revisions.

Once an item has been fully described in a schedule it should not be repeated at length. For example, having described the construction of a particular type of door, it can be given a type number or letter and this only would be repeated in the schedule. Characteristics, particularly on a finishes schedule, can often be defined by a code letter or number with an interpretation in the form of a key to one side. Constant repetition of descriptions is then eliminated. Well-recognized abbreviations which will not be confused with others can be used to save time. For example, 'KPS3' is readily recognized as 'knot, prime, stop and paint three coats of oil colour'.

Schedules are not intended to supplant the bills of quantities which will contain the full and final description of the characteristics of the items concerned.

Examples of the following schedules are given at the end of this chapter:

Example C   Window schedule including ironmongery, glazing, frames and fixings.

Example D   Door and ironmongery schedule including glazing, frames and thresholds.

Example E   Finishes and decorations schedule (sometimes including small standardized fittings such as hooks, hanging rails, etc.).

Example F   Schedules of manholes and covers.

This is by no means a complete and comprehensive list. There are many other items which may be described and collated conveniently in schedules but it is very difficult for them to be standardized as they vary so much with each different type of building.

The examples given vary from those shown in BS 1192:1984 but they are more easily understood and less likely to lead to error.

Drawings and Schedules for Records

Although reference has already been made in this chapter to the keeping of record drawings, it is a matter of sufficient importance to justify emphasis and clarification.

Before starting work on site there are two stages at which it is important to keep record drawings and schedules before they are subjected to any further amendment. These are:

(1) Documents upon which the bills of quantities are based – a full set of completed drawings and schedules (see Examples C, D, E and F).
(2) Contract drawings – those to which the contract documents refer; and marked accordingly.

It is useful for these copies to be kept in such form as permits them to be reproduced should any future disagreement arise.

True and accurate 'as built' drawings are of vital importance in the future maintenance of the building, particularly of the drainage, and copies should be made and issued to the employer upon completion of the contract. Their preparation depends upon regular correction of the drawings as the work proceeds and as amendments are issued. They may be in plastic negative form so that they are easily reproduced and read, or in microfilm form to minimize on storage space, but available for enlargement should the need arise.

EXAMPLE C: WINDOW SCHEDULE

WINDOWS

WINDOW TYPE REF:	A	B	C	D	E
WINDOW REF NO.	W2 to W19 inc. W22 to W39 inc.	W1, W20 W21, W40	W41, W42 and WW3	W44 to W50 inc.	W51
NUMBER OFF	36	4	3	7	1
MATERIAL OF MAIN FRAME	Extruded aluminium PVF2 treated, insulated sections: by Glazing International Limited				
SUB-FRAME	Nil	Nil	Nil	Nil	Hardwood
BUILDERS OPENING SIZE	Continuous horizontal opening height 1200		600Hx1500W	1800Hx750W	1800 diam
WINDOW UNIT SIZE	1185Hx1450W (inc cill)	1185Hx750W (inc. cill)	585Hx1485W	1785Hx735W	1485 diam
FIXING JAMB	To coupling mullion	Alum. brackets to brickwork	Aluminium brackets to brick		
HEAD	Alumium brackets plugged to concrete lintals or beam				Nil
CILL	Aluminium cill with brackets plugged and screwed to blockwork				Nil
SEALANT	Brown polysulphide to heads and jambs				

IRONMONGERY & GLAZING

FASTENINGS	Alum. lever handle and locking latch	Nil	Aluminium level handle	Aluminium peg stay	Nil
OPENING LIGHTS HINGES/PIVOTS	Horizontal hung projecting opening gear	Nil	Top hung (hinges with frame)	Side hung (hinges with frame)	Nil
STAYS	As part of opening gear	Nil	Pair of friction stays	Nil	Nil
OPERATING GEAR	Nil	Nil	Nil	Nil	Nil
SECURITY	Safety catch in jamb	Nil	Nil	Nil	Nil
GLAZING	Double glazed units, outer glass 6mm Pickertons Antison Silver, 12mm void, inner glass 6mm clear float				
FIXING	Extruded neoprene gaskets – all windows pre glazed by manufacturer				
NOTES					See drawing No. D02/64
REVISIONS Ref Wind Date A WA/21 28/2/92	DIAGRAMS				

Project Title   SHOPS & OFFICES – NEW BRIDGE STREET – BORCHESTER

Architects: REED & SEYMORE   WINDOW SCHEDULE   NO.   456/9.1   REV.   A

EXAMPLE D: DOOR SCHEDULE

DOORS

DOOR TYPE REF:	A		B	C	D
DOOR REF NO.	1	2 to 8	1 to 4	1 to 8	1 to 15
LOCATION & NUMBER	Main entrance lobby ⁞ 1	Entrance hall & corridor 01 & 11 ⁞ 7	Main staircase ⁞ 4	Rooms 03, 04, 05 & 10 to 14 incl. ⁞ 8	Offices 06 to 010 & 115 to 120 ⁞ 15
TYPE & DESCRIPTION	HW framed single swing double door; 6mm toughened glass		Plyfaced solid core flush panel with GWPP glazed viewing panel		Plyfaced solid core flush panel
FIRE RATING	Nil	Nil	60/60	30/30	Nil
DOOR SIZE	2 @ 2040x726x46 with 12mm rebated styles		2040x826x46 and 2040x426x46	2040x826x46	
FINISH	Polyurathene sealed		Colour preservative treated and wax polished		
FRAME OVERALL SIZE (nominal) & SECTION	2375x1850, 2375x1500 ex 150x60 rebated 25mm kicking rail ex 200x36		2375x1350 ex 150x50 rebated 25mm	2100x900 ex 150x60 rebated 25mm	2100x900 125x32 plus stops
MATERIAL/FINISH	Hardwood sealed		Softwood primed and painted		
FAN LIGHT	6mm PP		6mm GWPP	Nil	Nil
SIDE LIGHT	6mm toughened glass	Nil	6mm GWPP	Nil	Nil
SWING					
IRONMONGERY	1 @ 726 726	7 @ 726 726	4 @ 826 726	6 2	7 8
HINGES		1½ pair SS		1 pair rising butts	1 pair
LOCKS & LATCHES	Mortice deadlock type Ref. type –	Mortice dead lock type –	Mortice dead lock type –	Mortice latch and dead lock type –	
HANDLES	1 pair "D" pattern pull handles type –		Lever handles type –		
KICK PLATES	2 pairs satin aluminium 200 high		Nil	Nil	Nil
PUSH PLATES	1 pair satin aluminium 200 high		Nil	Nil	Nil
BOLTS	2 off SA flush (to leaf without lock)		2 off SA flush to small leaf	Nil	Nil
STOPS & STAYS		Skirting mounted type –		Floor mounted door stop type –	
MISCELLANEOUS			Surface mounted overhead closers. Intumescent strip to 3 sides		
REVISIONS Ref ⁞Door ⁞Date A ⁞DA/2 ⁞28/2/92	DIAGRAMS				

Project Title   SHOPS & OFFICES – NEW BRIDGE STREET – BORCHESTER

Architects: REED & SEYMORE   DOOR SCHEDULE   NO. 456/13.2   REV.   A

EXAMPLE E: FINISHINGS SCHEDULE

FINISHES

FLOOR LEVEL	Ground	Ground	Ground	First	First
ROOM NAME	Entrance Lobby	Stair Lobby	Enquiries	General Office	Womens Lavatory
ROOM NO.	G01	G02	G03	I01	I02

FINISHES

SCREED & FLOOR FINISH	75mm screed Carpet, Swatch ZD 10016			Raised floor system with carpet inlays to trays	50mm screed 150x150 ceramic floor tile
WALLS 1	Glazed				Plaster
2	Sirpite plaster finish to sand and cement base coat				Glazed Wall Tiles
3					Glazed Wall Tiles
4					Glazed Wall Tiles
SKIRTINGS	4x100x19 HW as detail 456/W/26			19x100 SW	Flush, coved ceramic tile
CEILING	Concealed grid suspension system; Class 1 tiles		Lay-in tile suspension system: Class 0/1		Plaster board, scrim & set
MISC & NOTES	Standard brass mat well frame set into screed	Staircase see detail drawing 456/14.3	Reception counter see drawing 456/W/25		

DECORATIONS

WALL FINISHES	Vinyl wall fabric by Sampsons Ltd	1 mist coat plus 2 coats vinyl emulsion – matt			Prime plus 2 coats flat oil
COLOURS 1	Glazed Ref No. SAM 1234	BS 00A01	BS 20C37	BS 16C33	BS 18E51
2	Ref No. SAM 1234	BS 00A01	BS 20C33	BS 16C33	2,3&4 tiled
3	Ref No. SAM 1234	BS 00A01	BS 20C33	BS 16C33	
4	Ref No. SAM 1234	BS 00A01	BS 20C33	BS 16C33	
CEILINGS	Nil	Nil	Nil	Nil	Flat oil white
FRAMES & ARCHITRAVES	1 + 2 coat gloss oil brilliant white				H2 gloss oil BS 20C40
DOORS	Glazed alumin	HW veneer 2 coats satin polyeurathene seal			Plastic laminate faced
WINDOWS	Aluminium frames self finished throughout				
CILLS	1 + 2 coats gloss oil brilliant white				Tiled as walls
SKIRTINGS	1 + 2 coats gloss oil brilliant white				1 + 2 gloss oil BS 20C40

REVISIONS		MISC & NOTES				
Ref	Rm No.	Date				
A	G01	1/3/92		Handrails – 2 coats glass polyeurathen seal	Reception counter – American oak french polished	Plastic laminate cubicles & ducting, extent of wall tiling etc, see drawing 456/W/27
B	I02	16/3/92				

Project Title   SHOPS & OFFICES – NEW BRIDGE STREET – BORCHESTER

Architects: REED & SEYMORE   FINISHINGS SCHEDULE   NO. 456/14.1   REV. B

EXAMPLE F: MANHOLE SCHEDULE

MANHOLES & COVERS

MANHOLE REF:	FM1	FM2	FM3	FM4	FM5
INTERNAL SIZE	1125x825	900 diam	1050 diam	1050 diam	1350 reducing to 900 diam
CONSTRUCTION & MAKE	225mm brick to 150 conc base	Precast concrete chamber rings, cover slabs and sealing rings set in concrete backfill. 150mm insitu concrete base			
COVER SIZE	450x600	450x600			
COVER TYPE	Med duty double-sealed	Grade C Fig 7		Grade C Fig 5	
INVERT LEVEL	7.315	7.214	7.02	6.807	5.705
GROUND LEVEL	8.46	8.153	8.175	8.08	8.07
COVER LEVEL	8.46	8.382	8.175	8.382	8.382
MAIN CHANNEL SIZE	100mm	100mm	100mm	100/150mm	150mm
TYPE STRAIGHT (S) CURVED (C) TAPPERED (T)	S	C	S	S / T	S
BRANCH CHANNELS SIZE TO RH	2x100mmx90°		1x100mmx90° 1x100mmx135°	1x150mmx90°	
SIZES TO LH	1x100mmx135°	1x100mmx90°	1x100mmx135°	1x100mmx90° 1x100mmx135°	1x100mmx90°
INTERCEPTING TRAP	100mm				
F.A.I.					100mm
BACKDROP INLET					As detail drawing 071/C
STEP IRONS					C.I. into precast units
MISCELLANEOUS	DPC between head of brick and G.F slab		MH cover raised to paving level on brick upstand		
REVISIONS Ref MH No Date A FM4 28/2/92	DIAGRAMS (Internal MH)				

Project Title — SHOPS & OFFICES - NEW BRIDGE STREET - BORCHESTER

Architects: REED & SEYMORE   MANHOLES & COVER SCHEDULE   NO. 456/   REV. A

Chapter 5

Specifications

Quality of Work

By preparing good drawings and schedules, we have shown how the architect can convey most of the information about a proposed building first to the quantity surveyor and later to the contractor. However, there is information which, by its nature, must be given by written description. For example, the quality of cement to be used or the method of laying asphalt cannot be shown graphically but must be described in writing.

There are some background principles which should be borne in mind when writing specifications. The specification must be carefully written whether bills of quantities are to be used or not, as it is the basis for establishing the quality of work, and in some instances the quantity. It cannot be overstressed how important it is to set out the information in a clear, logical and unambiguous manner. Writing a specification is a difficult and often tedious task, for which few have had any formal training.

Traditional Methods

The usual process is to prepare a full specification (a 'traditional specification'), based on the same trade-by-trade or work sections basis as that used in the Standard Method of Measurement (SMM).

A traditional specification is often set out in two distinct sections. First come materials and workmanship clauses, which describe the quality of work and the type of materials to be used. These will generally be standard clauses which require little alteration for specific projects. Secondly, there will be the schedule of work, which sets out the extent of the work and which must be specifically prepared for each project. There will be differences, as discussed below, depending on whether there are to be bills of quantities or not.

If bills are required, probably the best procedure is for the architect to prepare notes as the design is developed and then for the quantity surveyor to go through them, making his own copy and dealing with any queries or points of clarification that may arise.

When completing the notes against the specification headings, matters already adequately dealt with on drawings or schedules should not be repeated, but a simple cross-reference given. The notes should be brief and to the point, and long descriptions can often be avoided by the

EXAMPLE G: SEQUENCE SPECIFICATION

(1) Remove gas fire and gas pipe and clear out grate.
(2) Carefully remove marble fireplace surround and cast-iron fireplace and set aside for removal by client.
(3) Remove hearth and debris to base of existing sleeper walls.

Where hearth is removed:
(4) Build new sleeper walls to line through with existing.
(5) Install new treated timber wall-plate on dpc type X.
(6) Install new floor boards.
(7) Brick up existing fireplace opening, with new dpc to lap over existing and new plaster ventilator 200 mm above floor level.
(8) Remove defective plaster, prepare surfaces and apply new plaster (mix Y).
(9) Install new skirting (pattern Z) across opening.
(10) For decorations, see schedule ref. . . .

---

inclusion of sketches. The materials and workmanship clauses will be incorporated into the preambles section of the bills.

As the vast majority of projects are arranged on a trade basis, the specification headings listed below follow this pattern. From time to time, however, an alternative arrangement may be preferable. Particularly on refurbishment work, it can be easier to describe the work on a room-by-room basis. It is far easier to describe the chronological sequence of work where, for instance, a fireplace is being adapted, a new window formed, or a doorway blocked up. A typical example is shown in Example G. It can be seen that most trades are involved in this operation; to split it into a trade-by-trade basis can involve complex cross-referencing quite out of proportion to the scale of the work. It is also much more convenient for those on site to be able to envisage the whole sequence of work.

At the other extreme, on projects of great complexity, perhaps where a management contractor is involved, a traditional trade-by-trade basis may again not be appropriate. If, for instance, a great number of specialist sub-contractors are involved, the SMM sub-divisions are inadequate to describe the work properly.

Other Methods of Specifying

In some industries, although limited in the construction industry, it is possible to use a reference specification. Here, the specification makes regular reference to other nationally accepted standards which are referred to by name only, and detailed description need not therefore be included. Clearly, this reduces the bulk of any individual specification, and makes it much quicker and simpler to produce. Against this, however, must be set the disadvantage that the nationally accepted standards must be available, and their content known in detail, by all those involved. Every construction industry specification moves to some extent towards a reference specification because it will mention British Standards without writing out their text in full. Very often, the specifier is unfamiliar with the full text of a BS, and the labourer on site tends to know even less. There has been much debate about how much text of any particular BS should be referred to in a specification, and there is no simple answer. Whenever a BS is referred to, it should be supported whenever possible with concise details relating to the project in hand, particularly where the BS gives alternative solutions.

BS 8000 (workmanship on building sites) has now been published in 15 parts, and is intended to be a reference specification for the construction industry. Each part covers one or more of the trades sections of the common arrangement of work sections of CPI, and it is hoped that BS 8000 will become a familiar document both to the design team and to contractors. As well as standard specification clauses, it contains a descriptive commentary and diagrams.

A further type of specification is the performance specification. As its name suggests, the intention is to set out the required performance, without setting out the details of how this is to be achieved. A simple example is where a contractor is required to install a heating system which is to achieve 21° C under certain specified conditions; the size and lay-

Where bills are used . . . the specification is not a contract document

out of radiators are for the contractor to decide. In this solution, as the contractor is taking responsibility for design, JCT 80 will need the addition of the Contractor's Designed Portion Supplement 1981 Edition.

The Specification as a Contract Document

Whichever form of specification is adopted, a further crucial point has to be established. If bills of quantities are used, the specification should be incorporated in the bill, and the form of contract will be the 'with quantities' version. However, if the scale of the project is such that bills are considered unnecessary, the form of contract will be the 'without quantities' version, and the specification then becomes a contract document. On smaller projects, a bill of quantities may well be perfectly adequate. In this case, the specification must include full details of contract particulars, preliminary items, and sufficient details to enable the contractor to price the work properly.

If, however, bills of quantities are to be prepared, the quantity surveyor will incorporate the architect's specification into the bills. It is essential, therefore, that the architect provides clear instructions to the quantity surveyor.

The following specification headings should provide sufficient information either to be incorporated in the bills of quantities or to be passed direct to the contractor.

Specification Headings

The common arrangement headings of CPI are shown in brackets

Tendering Particulars
(A – Preliminaries/general conditions)
(1) Name of job.
(2) Name, address, telephone number and reference of:
    (a) Employer.
    (b) Architect.
    (c) Quantity surveyor.
    (d) Structural engineer.
    (e) Service engineer.
    (f) Other consultants.
(3) Tendering details:
    (a) Date to be sent out.
    (b) Date and time of submission.
    (c) Where tenders to be submitted.
(4) List of contractors invited to tender.

Preliminaries
(A – Preliminaries/general conditions)
(1) Address of site.
(2) Place where all drawings may be inspected.
(3) Access for contractors to inspect site.
(4) Scope of contract.

(5) Procedure; any special conditions or sequence of operations.
(6) Particulars of access to site.
(7) Form of contract to be used.
(8) Any amendment or deletion in full.
(9) List of drawings for preparation of bills of quantities.
(10) Whether the contract is to be under hand or as a deed.
(11) Contract Appendix items to be completed:

Note: the following arrangement is taken from the JCT Standard Form of Building Contract, Private with Quantities 1980 Edition, incorporating amendments 1 to 9.

(a) Statutory tax deduction scheme (fourth recital and clause 31)
(b) Base date (clause 1.3)
(c) Date for completion (clause 1.3)
(d) VAT agreement (clause 15.2)
(e) Defects liability period (clause 17.2)
(f) Assignment by employer of benefits after practical completion (clause 19.1.2)
(g) Insurance cover for any one occurrence or series of occurrences arising out of one event (clause 21.1.1)
(h) Insurance – liability of employer (clause 21.2.1)
(i) Insurance of the works – alternative clauses (clause 22.1)
(j) Percentage to cover professional fees (clauses 22A, 22B.1, 22C.2)
(k) Annual renewal date of insurance as supplied by contractor (clause 22A.3.1)
(l) Insurance for employer's loss of liquidated damages – clause 25.4.3 (clauses 22D and 22D.2)
(m) Date of possession (clause 23.1.1)
(n) Deferment of the date of possession (clauses 23.1.2, 25.4.13, 26.1)
(o) Liquidated and ascertained damages (clause 24.2)
(p) Period of delay (clause 28.1.3)
(q) Period of delay (clauses 28A.1.1 and 1.3)
(r) Period of delay (clause 28A.1.2)
(s) Period of interim certificates (clause 30.1.3)
(t) Retention percentage (if less than 5%) (clause 30.4.1.1)
(u) Work reserved for nominated sub-contractors for which the contractor desires to tender (clause 35.2)
(v) Fluctuations (clause 37)
(w) Percentage addition (clauses 38.7 or 39.8)
(x) Formula rules) (clause 40.1.1.1)
(y) Settlement of disputes – Arbitration (clauses 41.1 and 41.2)

Note: the insurance provisions (clauses 20 to 23) are particularly complex, and it is essential that full reference is made to the Form of Contract and other supporting documents (for instance, JCT Practice Note 22 and Guide to Insurance Provisions).

(12) Local authority fees.
(13) Particulars of any other special conditions not normally anticipated.
(14) Temporary roads.
(15) Temporary buildings:
  (a) For contractor.
  (b) For architect and others.
  (c) For clerk of works.
  (d) Rates on foregoing.
(16) Temporary telephones:
  (a) Particulars.
  (b) Cost of employer's calls.
(17) Temporary screens.
(18) Temporary hoardings and gantries, and advertising rights on same.
(19) Any special security arrangements.
(20) Provision for supplying of samples.
(21) Provision for testing materials.
(22) Provision for fuel for drying out building.
(23) Provision for removing water below water table.
(24) Particulars of other works expected to be in progress on or adjacent to the site.
(25) Contingency.

Demolitions and Alterations
(C – Demolition/alteration/ renovation)
(1) General particulars.
(2) Old materials to be retained by employer.
(3) Old materials to be re-used.
(4) Temporary screens.
(5) Special shoring.
(6) Particular items of covering and protecting existing fittings, adjoining property.

Excavation and Earthwork
(D – Groundwork)
(1) Site preparation.
(2) Datum level.
(3) Site levels and floor levels.
(4) Nature of sub-soil.
(5) Trial holes and existing services.
(6) Level of sub-soil water and details of tide levels for river and sea.
(7) Return fill.
(8) Disposal of surplus soil.
(9) Hardcore or fill.
  (a) Type of material.
  (b) Maximum size of material.
  (c) Thickness of beds.
  (d) Weight of roller to be used.
  (e) Finishing for concrete or other surfacing.

Piling and Diaphragm Walling
(D – Groundwork)
Note: normally a consultant structural engineer will be employed for this item, and will supply a detailed specification.
(1) Nature of ground and strata.
(2) Type of piling giving full details.
(3) Level that piling is to commence.
(4) Precautions to be taken relating to adjoining properties.

(5) If piling is carried out by a specialist then builder's work items:
    (a) Setting out.
    (b) Disposal of pile excavations.
    (c) Cutting off pile caps.
    (d) Bending reinforcement at top of piles.

Concrete Work

(E – insitu concrete/large precast concrete)
Note: for larger or complex projects, the consultant structural engineer will supply the specification. The following items are intended for simple projects where a BS standards mix or nominal mix is specified.

(1) Cement.
(2) Fine aggregate.
(3) Coarse aggregate.
(4) Tests.
(5) Frost.
(6) Proportions of mix and where to be laid.
(7) Finish to formwork and where to be used.
(8) Reinforcement.
(9) Expansion joints and construction joints
(10) Damp proofing.
(11) Hardening and dust proofing.
(12) Key for plaster.
(13) Precast concrete.
(14) Hollow tile and precast concrete slabs.
(15) Prestressed concrete.
(16) Pavement lights.
(17) Column guards and masonry anchors.
(18) Building paper
(19) Contractor-designed construction giving full details.

Brickwork and Blockwork

(F – Masonry)

(1) Common bricks:
    (a) Below dpc.
    (b) Above dpc.
(2) Facing bricks:
    (a) Below dpc
    (b) Above dpc.
(3) Special bricks.
(4) Brickwork bonds.
(5) Mortar mixes.
(6) Pointing:
    (a) Facing bricks.
    (b) Fair-faced work internally.
(7) Ties in cavity walls: type and spacing.
(8) Damp-proof courses.
(9) Partitions:
    (a) Type of block.
    (b) Mortar mix.
(10) Decorative brickwork.
(11) Air bricks.
(12) Flue linings.
(13) Throat units.
(14) Chimney pots.
(15) Reinforcement.
(16) Method of closing cavities:
    (a) at jambs.
    (b) at cills.
(17) Vertical joints between blockwork and concrete.
(18) Vertical joints between blockwork and partition blocks.
(19) Horizontal joints between partitions and ceilings.
(20) Glass blocks:
    (a) Type and size.
    (b) Mortar mix.
    (c) Mastic.
    (d) Reinforcement.

(21) Gas flues.
(22) Fireplaces (including fires, boilers, stoves, surrounds).
(23) Slate and terrazzo cills and the like.

Underpinning
(D – Groundwork)
Note: normally a specialist firm or consultant structural engineer will be employed for this item and will supply a detailed specification.

(1) Description of structure to be underpinned giving:
    (a) Location.
    (b) Length on plan.
    (c) Depth of foundation below ground level.
(2) Description of new work giving:
    (a) Depth below existing foundation.
    (b) Limit of length of each operation.
    (c) Details of materials to be used including mix of concrete or type of brickwork.
    (d) Thickness of new walls.

Rubble Walling
(F – Masonry)
(1) Particulars of stone or quarry.
(2) Type of walling.
(3) Details of coursing.
(4) Finish.
(5) Mortars:
    (a) Bedding.
    (b) Pointing.

Masonry
(F – Masonry)

For each of the following:
(A) Natural stone (each type);

(B) Artificial and reconstructed stone;
state as applicable:
(1) Quarry.
(2) Bed.
(3) Mix.
(4) Aggregate.
(5) Colour.
(6) Texture and finish.
(7) Mortars:
    (a) Bedding.
    (b) Pointing.
(8) Bonding to backing.
(9) Treatment of back of stone.
(10) Cramps.
(11) Dowels.
(12) Joggles.
(13) Lengths of stones (sections as on drawings):
    (a) Copings.
    (b) Cills and string courses.
    (c) Mullions and jambs.
    (d) Columns and pillars.
(14) Protection.
(15) Sculpture and carving:
    (a) Where to be executed (on or off site).
    (b) Weight if executed off site.

Asphalt Work
(J – Waterproofing)
For each of the following:
(A) Horizontal tanking;
(B) Vertical tanking;
(C) Roof covering;
(D) Pavings;
state:
(1) British Standard.
(2) Thickness (and heights of skirtings).
(3) Number of layers.
(4) Colour.
(5) Finish.
(6) Felt underlay.
(7) Reinforcement.

(8) Skirtings and upstands to roof.
(9) Skirtings and upstands to pavings.

Roofing
(H – Cladding/Covering)
(1) For tiles, slates, state:
   (a) Description of materials.
   (b) Size.
   (c) Gauge.
   (d) Method of fixing.
   (e) Size of battens.
   (f) Finish at eaves.
   (g) Finish at verges.
   (h) Finish at abutments.
   (i) Ridges, including ends.
   (j) Hips, including ends.
   (k) Valleys.
   (l) Special fittings, glass tiles, vent tiles.
   (m) Underfelt.
(2) For metal and other corrugated coverings, state:
   (a) Description of material.
   (b) End and side laps.
   (c) Method of fixing.
   (d) Finish at eaves.
   (e) Finish at verges.
   (f) Finish at abutments.
   (g) Ridges, including ends.
   (h) Hips, including ends.
   (i) Valleys.
   (j) Expansion joints.
   (k) Special fittings, ventilators, roof lights.
(3) Roof decking:
   (a) Description of material giving kind of decking, thickness, quality and method of fixing.
   (b) Details of fixing blocks.
   (c) Details of all bearings, eaves, kerbs, flashings and nibs.

(4) Felt roofing:
   (a) Type of felt.
   (b) Number of layers.
   (c) Weight of each layer.
   (d) Bonding.
   (e) Applied finish.
   (f) Lapping and jointing.
   (g) Falls.
   (h) Finish at eaves.
   (i) Finish at verges.
   (j) Finish at abutments.
   (k) Treatment of ridges and hips.
   (l) Treatment of valleys.
   (m) Treatment of outlets.
   (n) Working around pipes, balusters.
(5) Sheet metal roofing:
For each of the following:
(A) Lead;
(B) Zinc;
(C) Copper;
(D) Aluminium ;
(E) Stainless steel;
state:
   (a) Weight or gauge.
   (b) Maximum size of sheets.
   (c) Methods of jointing and fixing.
   (d) Underlay.
   (e) Flashings.
   (f) Soakers.
   (g) Aprons.
   (h) Wedgings for flashings.
(6) Thatching:
   (a) Description of material giving kind of thatch, thickness, and method of fixing.
   (b) Finish at abutments.
   (c) Finish at eaves.
   (d) Finish at verge.
   (e) Finish at hips.
   (f) Finish at ridge.
   (g) Finish at valleys.

(h) Details of ornamental features.
(i) Description of wire netting.
(7) Sheet metal flashings and gutters giving full details.
(8) For all roofs, state:
    (a) Details of cross-ventilation.
    (b) Position of vapour barriers.

Woodwork

(G – Structural/carcassing metal/timber; H – Cladding/covering; K – Linings/sheathing/dry partitioning; L – Windows/doors/stairs; Z – Building fabric reference specifications)

Carcassing and First Fixings

(1) Type and/or stress grading of timber.
(2) Preservatives.
(3) Strutting for floor joists.
(4) Battens to edges of suspended ceilings.
(5) Noggings for plasterboard and other linings.
(6) Firrings.
(7) Insulation in floors, walls and roofs.
(8) Snowboards and gang boarding.
(9) Cistern casings.
(10) Flag staffs.
(11) Tie rods, straps.
(12) Bolts and other connectors.
(13) Ventilation requirements.
(14) Vapour barriers.

Second fixings

(1) Types of timber:
    (a) Softwoods.
    (b) Hardwoods.
    (c) Ply, blockboard, chipboard, hardboard.
(2) For each of the following:
(A) Boarded flooring;
(B) Strip flooring;
state:
    (a) Timber.
    (b) Jointing.
    (c) Margins.
    (d) Finish.
(3) Plain and matchboard linings.
(4) Panelled linings.
(5) Particulars of the following where not shown on detail drawings or schedules:
    (a) Doors and frames.
    (b) Windows and frames.
    (c) Borrowed lights.
    (d) Lantern lights.
    (e) Hatches.
    (f) Skirtings.
    (g) Cornices.
    (h) Friezes and dados.
    (i) Sub-frames.
    (j) Cupboard units.
    (k) Pelmets.
    (l) Cloak rails.
    (m) Pipe casings and access doors.
    (n) Blackboards.
    (o) Picture rails.
    (p) Architraves and cover fillets.
    (q) Window boards.
    (r) Staircases.
    (s) Shelving.
    (t) Trapdoors.
    (u) Other fittings.
(6) Information on the following where not shown on detail drawings or schedules:
    (a) Ironmongery for doors.
    (b) Ironmongery for windows.

(c) Ironmongery for fittings.
(d) Fixing cramps for joinery.
(e) Dowels for joinery.
(f) Water bars.
(g) Shelf brackets.
(h) Handrail brackets.
(i) Hat and coat hooks.
(j) Mats.
(k) Lettering and numerals.
(l) Curtain tracks.
(m) Special ironmongery.

Structural Steelwork
(G – Structural/carcassing metal/timber)
Note: details will normally be supplied by the consultant structural engineer; the architect must ensure that notes are included to cover general builder's work items, for instance:
(a) Preparation of steelwork on site.
(b) Protection.
(c) Site treatments and finishes.

Metalwork
(H – Cladding/Covering; L – Windows/doors/stairs; N – Furniture/equipment; P – Building fabric sundries; Z – Building fabric reference specification)
(1) Metal windows,doors and rooflights; state:
  (a) Material.
  (b) Section.
  (c) Fixing lugs.
  (d) Shop finish.
  (e) Glazing beads.
  (f) Ironmongery.
  (g) Gearing.
  (h) Curtain track brackets.
  (i) Bedding and pointing.
  (j) Sub-frames and cills.

(2) Curtain walling.
Note: great care must be taken to ensure that all the design requirements of the curtain walling system are included, particularly relating to joint, weather exclusion and loadings.
(3) Particulars of the following when not shown on detail drawings:
  (a) Railings and balustrades.
  (b) Staircases and balconies.
  (c) Cat ladders and gangways.
  (d) Fire doors.
  (e) Shutters.
  (f) Collapsible gates.
  (g) Other doors and gates.
  (h) Ventilators and grilles.
  (i) Shop fittings.
  (j) Steel partitions.
  (k) Blinds.
  (l) Refuse hoppers and fittings.
  (m) Mat frames.
  (n) Trim.
(4) Particular items of covering and protecting finished work.

Plumbing and Mechanical Engineering Installations
(R – Disposal systems; S – Piped supply systems; T – Mechanical heating/cooling/refrigeration systems; U – Ventilation/air conditioning systems; Y – Services reference specification)
(1) For each of the following:
(A) Rainwater gutters;
(B) Rainwater pipes;
state:
  (a) Material.
  (b) Size.
  (c) Type.

(d) Method of fixing.
(e) Method of jointing.
(f) Connection of pipes to
    drains.
(2) Roof outlets.
(3) Rainwater heads.
(4) Gratings.
(5) Provisional or PC sum to
    cover work by Statutory or
    local authority for:
    (a) Incoming mains water.
    (b) Surface water and foul
        water drainage.
(6) Pipes.
    For each of the following:
(A) External water main;
(B) Rising main;
(C) Cold water services;
(D) Hot water services;
(E) Overflows;
(F) Wastes;
(G) Soil pipes;
(H) Vent pipes;
(I) Expansion pipes;
state:
    (a) Material.
    (b) Method of jointing.
    (c) Method of fixing.
    (d) Type of fittings.
(7) Stop valves.
(8) Bib valves.
(9) Drain off cocks.
(10) Safety valves.
(11) Ball valves.
(12) Meters.
(13) Sanitary fittings:
    Give particulars or state name
    of supplier and reference to
    quotation and/or schedule.
(14) Traps:
    (a) Material.
    (b) Type.
(15) Cisterns and tanks.
    For (A) cold and (B) hot
    water state:

(a) Type and quality.
(b) Capacity.
(c) Size.
(16) Lagging and other
    insulation:
    (a) Pipes.
    (b) Cisterns.
(17) Ductwork.
Details for heating and ventilation
systems will usually come from a
specialist consultant and, if a bill
of quantities is to be produced, full
details will have to be provided.

Electrical Installation and Specialist Services

(V – Electrical supply/power/light-
ing systems; W – Communication/
security/control systems; X –
Transport systems; Y – Services
reference specification)
Note
(1) It is *not* anticipated that the
    architect will normally
    prepare information from
    which the quantity surveyor
    can directly produce a bill
    of quantities for these specialist
    services, but rather that this
    will come from a consultant
    or alternatively that a PC
    sum will be included in the
    bills.

(2) The architect will wish to
    prepare some specification
    notes for the consultants.
    This will range from
    decisions on electrical fit-
    tings (particularly lighting)
    on simple schemes, to a
    detailed performance
    specification on complex
    projects

(3) The following information
    is intended only as

a guide to the quantity surveyor to the sources from which he is to expect his information.

For each of the following specialist services:

(A)   Electrical services;
(B)   Gas services;
(C)   Heating;
(D)   Hot water;
(E)   Ventilation;
(F)   Sprinklers and dry risers;
(G)   Lifts, escalators and conveyors;
(H)   Mechanical plant;
(I)   Kitchen equipment;
(J)   Lightning conductors;
(K)   Telephones;
(L)   Alarm and call systems;
(M)   Radio and television systems;
(N)   Earthing systems;
(O)   Provisional or PC sum to cover work by statutory authorities for electricity, gas and telephone mains services;

give name and address of consultant if any and state method by which each service will be dealt with in the bill:

(a)   Measured in bills of quantities.
(b)   PC from sub-contract bills of quantities.
(c)   PC from specification and drawings, or by selection.
(d)   PC as (c) but design developed by sub-contractor.
(e)   PC from artist or craftsman.
(f)   PC from local authority or public utility.

Work in Connection with Services

(P – Building fabric sundries)

Note:

Much of this work, particularly trenches, holes, chases and the like, will be self-explanatory from the drawings and particulars of the services involved prepared by the architect or his consultant, The following items, however, may require special mention:

(1)   Stop cock and valve pits:
    (a)   Sizes.
    (b)   Materials.
    (c)   Covers.
(2)   Bases for pumps, motors, boilers and other equipment:
    (a)   Sizes.
    (b)   Materials.
    (c)   Mortices.
    (d)   Special finishes.
    (e)   Ash pits.
(3)   For each of the following:
(A)   Ducts in floors;
(B)   Ducts in walls;
state:
    (a)   Sizes.
    (b)   Pipe bearers.
    (c)   Filling.
    (d)   Covers.
    (e)   Access traps.
(4)   Pipe casings:
    (a)   Sizes.
    (b)   Materials.
    (c)   Access panels.
(5)   Tank bearers.
(6)   Tank casings.
(7)   Pipe sleeves.
(8)   Hangers and brackets to be fixed by general contractor.
(9)   Soot doors.
(10)   Draught stabilizers.
(11)   Backboards for sanitary fittings and equipment.
(12)   Floor channels:
    (a)   Type.

(b) Size.
(c) Outlets.
(d) Covers.
(13) Bearers or frames for electrical switch gear.
(14) Cupboards for electric switch gear and meters.
(15) Fuel boards and baffles.
(16) Fuel for testing installation.
(17) Fittings when not included with installation:
    (a) Lighting fittings.
    (b) Cookers.
    (c) Refrigerators.
    (d) Wash boilers.
    (e) Water heaters.
    (f) Gas fires.
    (g) Other unit heaters.
    (h) Pumps.
    (i) Fuel bins.
    (j) Water softeners.
    (k) Machinery.
    (l) Kitchen equipment.
  (m) Fire-fighting equipment.
(18) Paint on pipes (if different from that specified in Painting and Decorating):
    (a) Cold water pipes.
    (b) Waste and soil pipes.
    (c) Hot water pipes.
    (d) Heating pipes.
    (e) Gas pipes.
    (f) Electrical conduit.
    (g) Other service pipes.
(19) Further works in connection with specialist services.

Floor, Wall and Ceiling Finishings
(M – Surface finishes)
(1) For each type of paving or floor covering state:
    (a) Description.
    (b) Mix.
    (c) British Standard.
    (d) Thickness and size of units.
    (e) Colour.
    (f) Screed or floated bed.
    (g) Bedding and jointing.
    (h) Adhesive.
    (i) Hardening, dust-proofing and sealing.
    (j) Polishing.
    (k) Non-slip finish.
    (l) Expansion joints and separation strips.
  (m) Skirtings to pavings.
(2) For each of the following:
(A) Block flooring;
(B) Parquet flooring;
state:
    (a) Timber.
    (b) Jointing.
    (c) Margins.
    (d) Finish.
(3) Metal lathing:
    (a) Gauge and mesh.
    (b) Surface finish.
    (c) Method of fixing.
(4) Plasterboard:
    (a) Type of board.
    (b) Method of fixing.
    (c) Jointing and scrimming.
    (d) Plaster finish.
(5) Acoustic tiles.
(6) Bonding agents.
(7) Ceiling and wall plaster:
    (a) Number of coats.
    (b) Type and mix of plaster for each coat.
    (c) Finish to salient angles.
(8) Coves and cornices, including bracketing.
(9) Fibrous plaster.
(10) Wall boards, type, thickness and fixing.
(11) Fabric and similar wall coverings, type and fixing.

(12) Staircase and step finishes:
(a) Treads.
(b) Risers.
(c) Strings.
(d) Skirtings.
(e) Nosings.
(f) Landings.
(13) Renderings:
(a) Types.
(b) Number of coats or thickness.
(c) Mix for each coat.
(d) Method of application.
(e) Finish.
(14) Cement and sand screeds:
(a) Mix.
(b) Thickness and falls.
(c) Waterproofing.
(15) Lightweight screeds:
(a) Aggregate.
(b) Mix.
(c) Thickness and falls.
(d) Toppings.
(16) Cement glaze.
(17) Terrazzo:
(a) Aggregate and matrix.
(b) Thickness.
(c) Mix.
(d) Finish.
(e) Separation strips.
(f) External angles.
(g) Internal angles.
(18) Wall tilings:
(a) Type and quality.
(b) Colour.
(c) Thickness and size.
(d) External angles.
(e) Internal angles.
(f) Bands.
(g) Method and type of bedding, jointing and pointing.
(19) Mosaic:
(a) Type.
(b) Colour.

(c) Thickness and size.
(d) External angles.
(e) Method and type of bedding, jointing and pointing.

Glazing
(L – Windows/doors/stairs)
(1) For each sort of glazing state:
(a) Type of glass.
(b) Weight or thickness.
(c) Quality.
(d) Method of glazing.
(2) Patent glazing.
(3) Roof lights.
(4) Mirrors and other glazing accessories.
(5) Particular items of covering and protecting finished work.

Painting and Decorating
(M – Surface finishes)
(1) For each of the following:
(A) Oil-based paint;
(B) Cement-based paint;
(C) Water-based paint (emulsion paint);
(D) Paint on metal externally;
(E) Paint on metal internally;
(F) Paint on galvanized metal externally;
(G) Paint on galvanized metal internally;
(H) Paint on wood externally;
(I) Paint on wood internally;
(J) Paint on other surfaces;
state:
(a) Name of manufacturer.
(b) Preparation of surfaces.
(c) Primer.
(d) Type of paint and number of undercoats.
(e) Type of paint and number of finishing coats.
Particulars of the following if required:

(2) Work in parti-colours.
(3) Staining.
(4) Wax polish.
(5) French polish.
(6) Lettering.
(7) Other types of finish.
(8) Paper-hanging:
    (a) Type of paper and PC.
    (b) Method of hanging.
    (c) Preparation.
(9) Particular items of covering and protecting finished work.

Drainage
(R – Disposal systems)
(1) For each of the following:
(A) Surface water drains;
(B) Soil drains;
(C) Drains for special effluents;
state:
    (a) Type and grade of pipe.
    (b) Jointing.
    (c) Particulars of bed or surround.
(2) Land drains.
(3) Gullies and traps:
    (a) Types.
    (b) Gratings.
    (c) Sealing plates.
    (d) Raising pieces.
    (e) Inlets.
    (f) Kerbs.
(4) Cleaning eyes.
(5) Connections to sewer:
    (a) To be done by.
    (b) Approximate cost.
(6) Manholes.
(7) Fresh air inlets.
(8) Interceptors for petrol and other special effluents.
(9) Septic tanks.
(10) Cesspools.
(11) Soakaways.
(12) Sewage disposal plants.

External Works and Fencing
(Q – Paving/planting/fencing/site furniture)
(1) For each of the following:
(A) Roads;
(B) Paths;
(C) Other pavings;
state:
    (a) Excavation or filling.
    (b) Base: material and thickness.
    (c) Finish: material and thickness.
    (d) Weight of roller for consolidation.
    (e) Expansion joints.
    (f) Kerbs, channels and edgings, spur stones and bollards, and road-marking.
(2) For each of the following:
(A) Boundary walls;
(B) Fences;
state:
    (a) Type.
    (b) Height.
    (c) Foundations or bases.
    (d) Piers or posts.
    (e) Filling between piers or posts.
(3) Gates.
(4) Crossovers.
(5) External signs.
(6) Road lighting:
    (a) Standards.
    (b) Fittings.
    (c) Bases for standards.
    (d) Trenches for cables.
    (e) Other work in connection.
(7) Laying service mains:
    (a) Water.
    (b) Gas.
    (c) Electricity.
(8) Other external work in connection with services.

(9) Soft landscaping: (a) Preparation. (b) Seeding and turfing. (c) Hedging and ditching.	(d) Schedule of trees and description of tree pits. (e) Schedule of plants.

Complexity of Building

The complexity of building is such that it is often impossible to state all the relevant information on the drawing. A number of cross-referencing systems have been devised to relate the drawings to the specification notes and the bills of quantities. Sometimes attempts have also been made to relate information within the wider context of the construction industry library: for instance, the CI/SfB system (see Chapter 4 under the heading Types, Sizes and Layout of Drawings). Other systems relate more closely to the production information stage, such as the National Building Specification.

The use of word processors has enabled complex descriptions to be held on disc, and then inserted in a particular specification only when required, without the need for re-typing or checking. Some very sophisticated systems now exist. For instance, that available from the National Building Specification allows technical reference information, standard specifications, and individual office specifications all to be interlinked. Furthermore, this system will soon be directly linked with drawings produced by CAD programs, to ensure that all information on drawings is accurately and comprehensively linked with the specification.

Europe

The European Community (EC) is also actively working on the process of harmonization. The 'single market' will involve the introduction of revised standards of workmanship, certification and technical methods. The pace of integration is now moving so rapidly that it is unwise to attempt to set down anything more than general guidelines. However, it is essential to realize that European standards are going to have a considerable influence throughout the industry.

The European Committee for Standardization (CEN) is working to co-ordinate national standards; products that comply are able to use the CE mark, although it is not yet clear how CE marks will harmonize with the British Standards Institution's kitemark. In principle, products carrying the CE mark may be sold anywhere in the EC without having to satisfy any additional technical requirements, although there are various sub-categories which may limit their use in some countries. The European Technical Approval is an alternative route for a manufacturer to obtain a CE mark, similar to the Agrément certificate procedure in the UK.

European standards (designated EN) and design codes are also rapidly being developed, and the intention is that when any European standard is published, an identical one will also be published in the UK with a BS number. The structural Eurocodes start with EC 1 (general principles) then describe particular materials (e.g. EC 5 covers timber structures) and are also to be extended to cover, for instance, fire precautions.

The Construction Products Directive came into force in the UK in December 1991, and as with all EC directives is intended to state general principles and objectives, the details being supplied by member states as necessary. The directive is essentially a performance specification and requires a product to be fit for its use; there are six 'essential requirements' (for instance, safety in case of fire), and products satisfying these requirements may qualify for a CE mark, currently it is understood that products accredited in a member country prior to June 1991 will also comply (at least during transitional periods) after that date; however, the intention is that there will eventually be harmonized standards for all European products and regulations. It is not yet agreed whether all products will eventually have to either carry a CE mark or be withdrawn.

Bills of Quantities

Tender and Contract Document

As in other fields of communication, new techniques and procedures of tender procurement are always being tried, with the traditional method using bills of quantities being subject to much criticism as to its effectiveness. However, it is the flexibility that can be brought to its composition and its inherent usefulness as a post contract monitoring tool that ensure its durability.

Bills of quantities are firstly tender procurement and secondly contract documents and as such have specific roles to fulfil. It is therefore important that they are prepared in accordance with the conditions of the contract, that they contain certain basic information, and that they are presented in a recognizable format which will facilitate their use.

Presented in a recognizable format

However, it has long been recognized that, in preparing this basic information, the quantity surveyor processes a great deal of detailed information, much of which could be made available to the contractor. This would be of use not only in tendering, but also in contract planning and administration. It is therefore always worthwhile, in the early stages of any job, spending some time considering the role the bills are required to play and what additional information could be of use to the particular parties involved.

The Role of Bills of Quantities

Although primarily designed as tendering documents, bills of quantities have an important contractual task in the pricing of variations. These variations, with the original contract sum, will form part of the final account. Additionally, bills of quantities are usually used in the computation of valuations for interim certificates.

There are, however, a number of additional roles that the bills can play. Of these the two most important are the locational identification of the work and the formation of a basis for cost planning, both of which will be discussed in more detail below.

Basic Information

The basic information contained in bills of quantities falls into three categories:

- Preliminaries;
- Preambles or descriptions of materials and workmanship;
- Measured works.

and within these three sections should be contained a complete description of the works and the conditions under which those works are to be carried out. In this way the bills of quantities become a co-ordinated whole and a complete financial representation of the works, in support of the contractor's tender.

The preliminaries should contain a definition of the scope of the works and details of the proposed form of contract, indicating the employer's intention in relation to all the options within the contract, as well as the details required to complete the appendix. All proposed amendments to the form of contracts should be set out in full, although it is recommended that the standard forms are used without amendment, whenever possible. The preliminaries should contain a detailed description of the administrative mechanisms that will be necessary to implement the conditions of contract and any special conditions as to the way in which the works are to be carried out. The preliminaries should also contain a list of the drawings upon which the documents are based and any special instructions in the method of pricing and presentation of the contractor's tender. The site and site conditions will need to be accurately defined and all restrictions relating to working methods clearly stated.

The preambles or descriptions of materials and workmanship are intended as a concise definition of all the materials required and of the standard or quality of workmanship to be employed in working or assembling them. This enables the measured work section to be kept reasonably brief and to be easily priced by contractor's estimators. These preambles must convey, either directly or by reference, the architect's and engineer's requirements as defined in their respective specifications.

The measured works section or sections, which can include fully measured mechanical and electrical services as well as building works, should form a detailed description of the works, presented in accordance with the rules of measurement laid down under the contract (usually the SMM). This enables the contractor's estimator to price each individual item of work according to recognised conventions. The measured works will usually also contain prime cost (PC) and provisional sums. It is important to recognize the difference between these two elements and to appreciate the procedures that are laid down for their use in the standard forms of contract (Chapter 7 describes in detail the procedures for sub-contactors and suppliers). For convenience a brief definition of each follows:

- *Prime cost sums* are used for works to be carried out by nominated sub-contractors and statutory authorities or for goods to be supplied by nominated suppliers, for which the sub-contractor will usually have been selected in the pre-contract stage and for which estimates or tenders should have been obtained.

  The JCT Standard Form of Nominated Sub-Contract Tender (NSC/T) is designed for this purpose (see Chapter 7). Provision will be made in the bills for the contractor to add his profit to these sums; in the case of sums for sub-contract work, items will also be included for general attendance and such items of special attendance as have been identified in the JCT standard form or in discussions with the particular specialists at the pre-contract stage. Sums for nominated suppliers will be accompanied by items for the 'fixing only' of the goods concerned.

- *Provisional sums* are used for work which cannot be fully detailed or for costs which are unknown at the time the bills are prepared.

  Under the rules of the SMM, these sums are now divided into sub-headings entitled 'defined' or 'undefined'. In the former case, where a greater idea of the work involved is known, the contractor must have allowed for the associated programme and time implications within his tender. Where the sums are 'undefined', such as for contingencies, the contractor is entitled to receive an allowance in his programme, together with payment for any associated additional preliminaries costs, for work ordered against such 'undefined' sums.

The architect must issue instructions, in accordance with the contract conditions, regarding the expenditure of these sums, which may cover measured builders' work, further nominated sub-contractors' works or nominated suppliers' goods. They must therefore be sufficient to cover the full cost of the work involved which, in the case of a nomination, includes the profit and general and special attendance items and also, where appropriate, additional preliminaries.

Additional Information

The publication of SMM7 in 1988 heralded the arrival of co-ordinated project information. CPI provides a structured method of reconciling the disparate elements of project information throughout the project documentation, specifications, drawings, bills of quantities and the like. The bills of quantities can therefore be directly related through CPIs numbering system to a definite clause in the architect's specification and to a particular detail on the architect's drawings. Most significant however, in practical terms, is the requirement to divide the measurement into sections that permit the bills to be more easily split up by the contractor for sub-contract tendering purposes.

The desire to make bills of quantities more useful usually leads to the provision of additional information which involves the retrieval or classification of the detailed information identified in the measurement process. The specific use required for the eventual bills will dictate the type and format of the extra information provided.

The complexity of projects has led to a need for some locational identification of the items in the bills and this has encouraged the development of annotated bills of various kinds whereas the increased use of cost planning techniques has led to the development of bills split into elements rather than work sections.

A brief description of these different documents follows and examples of the more common ones are set out at the end of this chapter. It is important to note that all the different formats are produced from the same basic measurement process and that, as long as the requirements are known before measurement starts, there is no reason why the information should not be presented in any of the alternative formats available. Indeed, as long as the correct coding is done during the original measurement, there is no reason why the quantities should not be produced in one format and later be re-sorted for an alternative presentation.

Computer processing of quantities can make this particularly easy.

Traditional and Modern Bill Formats

A bill produced under the SMM7s common arrangement sections will look like Example H with the quantities for each item presented in a single total irrespective of location. Section totals are collected in a summary to ascertain the total value of the works.

Additional information can be added to these presentations by providing extra columns in the bill for a location or elemental breakdown of the total quantities (Example J). With this extra breakdown the location or elemental cost totals can quickly be calculated for cost planning purposes.

If the cost planning function is of prime importance, the bills can be presented in elemental format (Example K). In this case the work is broken down into standard elements, normally those used by the Building Cost Information Service (BCIS) of The Royal Institution of Chartered Surveyors. When broken down in this way, the total quantities of any particular work section may not be immediately available, being split between elements, and this can be inconvenient for contractors' estimators, especially when obtaining sub-contract prices. However, with production of bills by computer, it is possible to overcome this difficulty by having the information appropriately coded and produced in two or even more formats: a work section bill for tendering, an elemental bill for cost planning and an operational bill for production control or contract administration.

Occasionally there is a need for an even more specialized format for the bills: for instance, where facilities for the analysis and control of labour resources are required, as in direct labour contracts. In these cases, an operational bill may be appropriate. The operational bill can be presented broadly in either traditional trade or elemental format, but with the difference that the work items are finally broken down into labour operations or activities. These operations or activities might best be likened to the breakdown that a contractor would produce for bonus payments to his work force.

If greater detail of the location of the billed items is required, the annotated bill can be used. Annotated bills can be prepared in several forms. The annotations can be provided as a separate document; they can be bound into the back of the bill; or they can be reproduced directly opposite each item in the document as shown in Example L. Annotations can be added to any of the formats previously described and can include reference to detailed drawings or schedules as well as amplified descriptions or simple locational references.

The management contract provides the employer with the expertise of the contractor in managing a contract from an early stage. The bills of quantities for this type of contract is split into as many 'packages' or 'work contracts' as are decided by the design team and the contractor. Each set of contract documents with its bill of quantities is a separate entity; whether the series of packages be set up on a 'work section' or 'elemental' basis, each package can comprise several trades and elements or only one. The contractor's advice is relied upon here to attain the most efficient method of structuring the scheme but care must be taken to ensure that the interfacing of the various packages is carefully examined to ensure items of general builders' work, for example, are not overlooked. Bills of quantities for construction management projects would follow the same principles.

Changes can occur in the presentation of bills of quantities according to the needs of the employer, the professional team and contractor on the project. It is up to the professional team to decide how best the flexibility of the system can be used to the benefit of all concerned, not forgetting that efficient documentation should lead to efficient working and hopefully to some cost advantage to the employer.

The following examples illustrate how bills of quantities can be produced in different formats to give costing and specification information to assist the contractor, the quantity surveyor and the architect.

The examples are not comprehensive and the format of bills of quantities can be adapted to suit each project in order to provide the cost information in the most advantageous manner.

EXAMPLE H: COMMON ARRANGEMENT BILL

Note: Extracts from three separate sections are set out in order to allow comparison with other formats. The order of the sections, and the items within them, generally follows the layout of the SMM.

				£
	F10 Brick/Block walling			
	Lightweight concrete blockwork to BS 6073			
	Walls			
A	100 mm thick	427	m²	
B	100 mm thick facework one side	78	m²	
C	100 mm thick curved on plan 900 mm			
	radius	12	m²	
D	200 mm thick	219	m²	
	M20 Plastered coatings			
	14 mm thick plaster to BS 1191 Part 2			
	comprising 12 mm thick undercoat and			
	2 mm thick finishing coat steel trowelled			
	Walls			
H	over 300 mm wide; to brickwork	860	m²	
J	over 300 mm wide; to blockwork	2168	m²	
	Ceilings			
K	over 300 mm wide; to concrete	1435	m²	
	Isolated columns			
L	not exceeding 300 mm wide	22	m²	
	M60 Painting/clear finishing			
	Painting plaster; prepare one mist and			
	two full coats of emulsion paint			
	General surfaces			
T	over 300 mm girth	4485	m²	
			£	

EXAMPLE J LOCATION BREAKDOWN BILL

Note: Extracts from two formats are set out. In the first example, the breakdown into categories A, B C could be used to identify different house types, buildings of phases. In the second example, the breakdown is into the BCIS standard list of elements and their references are used (see chapter 3).

	F10 Brick/Block walling				£
	Lightweight concrete blockwork to BS 6073				
	Walls				
A	100 mm thick	A 340			
		B 67	427	m²	
		C 20			
B	100 mm thick facework				
	one side	A 45			
		B —	78	m²	
		C 33			
C	100 mm thick curved on				
	plan 900 mm radius	A —			
		B 3	12	m²	
		C 9			
D	200 mm thick	A 99			
		B 57	219	m²	
		C 63			

	F10 Brick/Block walling				
	Lightweight concrete blockwork to BS 6073				
	Walls				
A	100 mm thick	2G	427	m²	
B	100 mm thick facework				
	one side	2E	78	m²	
C	100 mm thick curved on plan				
	900 mm radius	2D	12	m²	
D	200 mm thick	2G	219	m²	

£

EXAMPLE K: ELEMENTAL FORMAT BILL

Note: Extracts from two elements are set out, the BCIS element references being quoted. The Common Arrangement headings appear, as appropriate, as subsidiary headings, under the main elemental sections.

				£
	INTERNAL WALLS AND PARTITIONS (ELEMENT 2G)			
	F10 Brick/Block walling *Lightweight concrete blockwork to BS 6073*			
	Walls			
A	100 mm thick	427	m²	
B	100 mm thick facework one side	78	m²	
C	100 mm thick curved on plan 900 mm radius	12	m²	
D	200 mm thick	219	m²	
	WALL FINISHES (ELEMENT 3A)			
	M20 Plastered coatings *14 mm thick plaster to BS 1191 Part 2 comprising 12 mm thick undercoat and 2 mm thick finishing coat; steel trowelled*			
	Walls			
H	over 300 mm wide; to brickwork	860	m²	
J	over 300 mm wide; to blockwork	2168	m²	
	M60 Painting/clear finishing *Painting plaster; prepare, one mist and two full coats of emulsion paint*			
	General surfaces			
T	over 300 mm girth	3028	m²	
				£

EXAMPLE L: ANNOTATED BILL

Note: Annotations are best set out on the facing page opposite the items they amplify.

Annotations

Item
A   Non-load bearing partitions first floor (drwg 456/78).
B   Internal skin of hollow walls to plant rooms etc.
C   Stores ground and first floor and stair case enclosure walls.

D   Non-load bearing partitions ground floor (drwg 456/77).

H   Load bearing partitions ground floor.
J   Internal skin of hollow walls, first floor partitions and external face of store walls.
K   Soffit of first floor.

L   Columns of hall.

T   Refer to colour schedule.

£

	F10 Brick/Block walling *Lightweight concrete blockwork to BS 6073*		
	Walls		
A	100 mm thick	427	m²
B	100 mm thick facework one side	78	m²
C	100 mm thick curved on plan 900 mm radius	12	m²
D	200 mm thick	219	m²
	M20 Plastered coatings *14 mm thick plaster to BS 1191 Part 2 comprising 12 mm thick undercoat and 2 mm thick finishing coat; steel trowelled*		
	Walls		
H	over 300 mm wide; to brickwork	860	m²
J	over 300 mm wide; to blockwork	2168	m²
	Ceilings		
K	over 300 mm wide; to concrete	1435	m²
	Isolated columns		
L	not exceeding 300 mm wide	22	m²
	M60 Painting/Clear finishing *Painting plaster; prepare, one mist and two full coats of emulsion paint*		
	General surfaces		
T	over 300 mm girth	4485	m²

£

Chapter 7

Sub-Contractors and Suppliers

In this chapter, we examine the sub-letting of building work from the practical and contractual points of view. At the end, there are brief sections on nominated suppliers, obtaining tenders and sub-contractor design.

Sub-Contractors

JCT contracts are generally drafted on the premise that the contractor appointed will carry out and complete the works; only with written consent may the contractor sub-let any of the work. However, sub-letting is customary in the industry and the contract conditions make full provision for such arrangements. Therefore consent so to do may not be unreasonably withheld.

It is also customary in the industry for certain sub-contractors to be 'nominated' or 'named', that is to be selected at the sole discretion of the employer or his professional advisers, and in effect imposed on the contractor. This arrangement requires special handling, owing to the restrictions placed on the contractor's freedom of choice, and again contractual conditions are laid down to cover the situation.

The arguments for and against nomination have been aired by many experts over the years and a brief review of the significant points is set out below. Suffice it to say that, as buildings have become more complex, the need for nomination has grown and, in parallel, so have the problems arising from the process. It is for this reason that the JCT has more recently incorporated considerable administrative detail in its contract documentation and has introduced the concept of naming as well as nomination.

These terms, and the procedures involved, will be explained below, after a review of the basic concept of the employer reserving to himself the right to select specific sub-contractors.

Specific Selection of Sub-Contractors (Nomination or Naming)

Construction has become more complex and early decisions have to be made on various elements of the building before the architect can proceed with the detailed design. Steel and concrete structural frames, for example, must be worked out in considerable detail before the architect can design the cladding of them; and the cladding itself may be a proprietary system. Mechanical and electrical services, which are frequently extremely complex and often together account for 30–50% of the cost of the building, make considerable demands on space for plant rooms and ducting, and the architect must know just what is required before he can finalize the storey heights and floor layouts.

Thus it is essential for decisions on a number of fundamental elements of the building to be made very early on in the design and, if beyond the consultants' briefs, this may make it necessary to select the principal specialist sub-contractors at the beginning of the design process. Sub-contractors involved in this way are often responsible for some of the detailed design. This frequently happens in the case of structural steel-work, mechanical and electrical installations and lifts.

This early involvement puts these sub-contractors into a special relationship with the employer and in due course creates the situation in which contractor and sub-contractor are required to enter into a contract with each other. The contractual arrangements at this stage require very careful attention.

Early standard forms of contract made special provisions for sub-contractors nominated in this way, but the provisions made had many short-comings, both in terms of contractual relationships and in the responsibility for the design elements. Disputes leading to arbitration or litigation involving nominated sub-contractors have been common. In addition the system has been abused by using nominations by way of prime cost sums in bills of quantities to cover sections of the work which have not been properly designed in the pre-contract stage. Contractors have for a long time been justified in regarding a bill of quantities containing a large number of prime cost sums as a portent of inefficient and uneconomic building. However, the new rules incorporated in SMM7, for the use of provisional and prime cost sums, should end this particular malpractice.

While it is common practice for contractors to sub-let substantial parts of the works to sub-contractors of their own choice, they see nominated sub-contractors as being imposed upon them and creating a number of problems. While there is an element of truth in this, it is not uncommon for contractors to try to blame the nomination system for some of their own shortcomings. However, nomination does place specific duties and responsibilities on the employer, the architect, the contractor and the sub-contractor and, if problems are to be avoided, each must co-operate with and respect the others. Each party should perform their designated roles for the benefit of the job as a whole and to the satisfaction of the employer at the end of the day.

The 1980 edition of the Standard Form of Contract (JCT 80) intro-

duced the basic and alternative methods of nomination with full supporting documentation. Subsequently, these two procedures have been superseded by the *1991 procedure* introduced by amendment 10 to JCT 80. In
the interim, the Intermediate Form of Contract (IFC 84) introduced a different concept, that of named sub-contractors.

It must be emphasized that these systems are not alternative procedures to be selected at will in the course of a project. They are procedures specific to the particular forms of contract and must be operated as
appropriate depending on the form of contract chosen.

In this book, we concentrate on JCT 80. This form imposes procedures and disciplines on architects and quantity surveyors in relation to
nominations, even in the pre-contract stage. The object is to sort out,
before the contractor and sub-contractor are instructed to enter into contract, most matters which could lead to dispute and delay.

The JCT 80 Basic Method of Nomination

Although now superseded by the *1991 procedure* (amendment 10), it
will be helpful to review the original system so as to understand the
development of the latest system and to deal with contracts still using it.

JCT 80 identifies four types of arrangement under which work may be
sub-let to a sub-contractor. Two of these relate to domestic sub-contractors and two to nominated sub-contractors.

A *domestic sub-contractor* may be:

● A sub-contractor to whom the contractor sub-lets part of the work, at
  his discretion, subject only to the consent of the architect;
● A sub-contractor selected at the sole discretion of the contractor
  from a list of not less than three firms named in the bills of quantities.

A *nominated sub-contractor* is defined as a sub-contractor whose selection is reserved to the architect, and who may be:

● Nominated by way of a prime cost sum (to be identified at a later
  date);
● Specifically identified by name.

This selection could be made in the bills of quantities or in an instruction
regarding the expenditure of a provisional sum or, subject to certain
qualifications, in a variation order or by special agreement between the
parties.

The basic method of nomination involved the full documentation which accompanied the Standard Form, namely:

- Tender NSC/1   The form of tender and agreement.
- Agreement NSC/2   The employer/nominated sub-contractor
                      form warranty.
- Nomination NSC/3   The instruction to the contractor.
- Sub-Contract NSC/4 The sub-contract conditions.

Once the decision to nominate using the basic method had been taken, the procedure laid down in the contract had to be followed and it was obviously sensible to adopt it when the first enquiry was made to the prospective sub-contractor. (There is no point in obtaining a preliminary estimate from the sub-contractor and later requesting a quotation using the proper documentation and finding that the quotation has risen substantially because the earlier estimate had not allowed for the various conditions and obligations required by the sub-contract.) Even for the initial enquiry, therefore, Tender NSC/1 and Agreement NSC/2 would be used as these documents initiated a dialogue which, by stages, resolved all the necessary details in a logical order, thus ensuring a fair price for a properly defined scope of work.

Tender NSC/1 contained the tender itself followed by Schedule 1, into which were inserted the particulars of the main contract and the sub-contract; two appendices dealing with fluctuations; and Schedule 2 into which were inserted particular conditions relating to the sub-contract.

The procedure for making a nomination under the basic method involved the circulation of the necessary forms between the architect, the sub-contractor, the employer and the main contractor, so that the various sections of the forms were completed in sequence, until the whole process was complete. Unfortunately this could become a very complicated and lengthy process and it was not popular, particularly with architects, who saw the whole process as being far too complicated.

Under the basic method there was full consultation between contractor and sub-contractor prior to nomination. The numerous matters which had hitherto been constant sources of trouble, particularly in relation to the programme of works and special attendance provisions, which were set out in Schedule 2 on NSC/1, were discussed and agreed between contractor and sub-contractor before the architect issued his instruction making the nomination.

By having Agreement NSC/2 executed by the sub-contractor and the employer when Tender NSC/1 was first submitted, the sub-contractor's responsibility direct to the employer with regard to the design of the sub-contract works, and the selection by the sub-contractor of goods and materials in connection with them, was established at the earliest possible moment.

The JCT80 Alternative Method of Nomination

Under the alternative method, also now superseded by the *1991 procedure*, Tender NSC/1 and Nomination NSC/3 were dispensed with and Agreement NSC/2a and Sub-Contract NSC/4a were the only documents used. These forms, denoted by the 'a' suffix, were the basic documents adapted to account for not using the other forms.

When the alternative method was to be used this had to be stated in the bills of quantities or in the architect's instruction under which the nomination was made. The bills or the instruction also had to state whether or not Agreement NSC/2a would be used, this being optional under the alternative method.

In effect, the alternative method was analogous to the common practice under earlier forms of the contract whereby the architect obtained a quotation from a sub-contractor and subsequently nominated that sub-contractor with no formal documentation. Under JCT 80, however, even using the alternative method, once the nomination was made the contractor and the sub-contractor had to enter into the Standard Form of Sub-contract NSC/4a. The main difference between NSC/4 and NSC/4a was that the latter incorporated an appendix into which were inserted the various matters which under the basic method were set out in Tender NSC/1. Under the alternative method, however, these matters would not have been agreed between the contractor and the sub-contractor before the nomination was made and the advantages stemming from prior agreement would have been lost.

As far as the use of Agreement NSC/2a was concerned (the employer/sub-contractor agreement under the alternative method) this would depend on the nature of the sub-contract works. Where the proposed sub-contractor was responsible for design or the selection of materials or goods then clearly Agreement NSC/2a would be used, especially as under JCT80 it is expressly stated that the contractor has no design responsibility in connection with the sub-contract work. In addition Agreement NSC/2a would also be used if the nature of the sub-contract works was such that delay or default by the sub-contractor might disrupt the works as a whole, giving the contractor the right to seek an extension of time.

Choice of Method (Basic or Alternative)

The choice of the method of nomination had to receive careful consideration during the pre-contract period. Except where proposed sub-contract works were covered by a provisional sum the decision on the method of making each nomination had to be taken before the bills of quantities were completed as the method had to be stated in the bills.

The basic method should be used in respect of the following:

- Any sub-contract in which design information had to be obtained from the sub-contractor during the pre-contract period.
- Any sub-contract, the performance of which was likely to affect the progress of the contract as a whole.
- Any sub-contract which required prior agreement between the contractor and the sub-contractor on matters of programme, performance and attendance.

Both the basic method and the alternative method could be used on the same main contract. Furthermore, the architect could substitute the basic method for the alternative method and vice versa during the course of the contract, provided he had not already issued a notice of nomination under the basic method or a nomination instruction under the alternative method. Such a substitution, however, was treated as a variation and the main contractor would be entitled to reimbursement for any loss or expense he might incur as a result of the substitution. It was clearly advisable, therefore, that positive decisions on nomination were made during the pre-contract period.

The 1991 Procedure for Nomination

JCT 80 Amendment 10 sets out the detailed procedure, known as the *1991 procedure* (to distinguish it from previous methods) and defines the documentation to be used for nominating sub-contractors. The basic definitions of a nominated sub-contractor and the contractual relationships are not changed by this amendment; only the procedure is changed. Further, as with the original JCT 80 procedures, the *1991 procedure* must be carefully followed.

The new documentation is as follows.

- NSC/T   Tender and Agreement, comprising:
  - Part 1   The Architect's invitation to tender;
  - Part 2   The tender by a sub-contractor;
  - Part 3   Particular conditions (to be agreed by the contractor and nominated sub-contractor).
- NSC/N   The architect's nomination instruction.
- NSC/A   The articles of nominated sub-contract agreement.
- NSC/C   The nominated sub-contract conditions (which are incorporated by reference into Agreement NSC/A).
- NSC/W   The employer/nominated sub-contractor agreement or warranty.

The procedure using these documents is simplified, compared with the basic method, and now comprises four stages, as follows.

(1)   The architect issues his enquiry to tenderers enclosing:
   (a) NSC/T with Part 1 completed
   (b) The drawings and documents describing the works (including specification and quantities, if appropriate)
   (c) The main contract appendix, completed as far as possible
   (d) NSC/W with page 1 completed
(2)   The sub-contractors complete and submit their tenders on part 2 of NSC/T and return NSC/W completed.
(3)   The architect selects the appropriate sub-contractor, arranges for the employer to approve the tender and execute NSC/W and issues the nomination NSC/N to the contractor (with a copy of NSC/W and NSC/T with Parts 1 and 2 completed).
(4)   The contractor and sub-contractor agree the particular matters in Part 3 of NSC/T and enter into the sub-contract agreement NSC/A, which should be done within ten days of the contractor receiving the nomination.

It should be noted that, since the contractor and sub-contractor have not discussed the matter until stage 4, there may be matters in Part 3 of NSC/T which are not immediately agreeable in the ten days allowed, so the new procedure lays down a mechanism for resolving the situation.

The following points should also be noted.

● NSC/T requires the insertion of the main contract completion date and level of liquidated and ascertained damages, so that the sub-contractor cannot plead ignorance of them or that they are too remote to affect him.
● NSC/W provides a guaranteed procedure for direct payment of a nominated sub-contractor where the contractor does not provide proof of discharge of payment to the architect.
● The new procedure allows for early final payment to the nominated sub-contractor when the architect certifies that practical completion of his works has been achieved.

This procedure is therefore much simpler and shorter than the basic method originally promulgated under JCT 80. However certain problems can arise, of which the following are examples.

If the contractor and sub-contractor cannot conclude an agreement in the ten days prescribed, then the contractor must give written notice to the architect, who must resolve the situation within a reasonable time by:

● Giving more time; or

- Instructing the contractor to comply with the nomination instruction, if he considers the problem to be of no significance; or
- Notifying the contractor that he recognizes the problem and resolves it by issuing further instructions so that the contractor can complete the agreement, or cancelling the nomination instruction and either omitting the work or issuing another nomination instruction. In this situation, the architect's instructions are a relevant event in relation to extensions of time and could give rise to a claim for loss and expense.

The contractor is entitled, within seven days of receiving a nomination instruction, to raise a reasonable objection to the nomination of a sub-contractor. If this is accepted by the architect, he must issue further instructions, which could lead to an extension of time and a claim for loss and expense.

Similarly, the sub-contractor may raise objections during the process. If he is not told who the contractor is at the start he may, within seven days of being told, withdraw his tender without giving any reasons, leaving the employer with no means of redress; further, of course, the contractor may again be able to claim an extension of time and loss and expense when the subsequent instruction is issued.

Naming under the Intermediate Form (IFC 84)

The Intermediate Form of Contract does not generally contain as much procedural detail as the Standard Form. There is no detailed definition of domestic sub-contractors and sub-letting as in the JCT 80 Form, and there is only a general reference to sub-contracting which shall not be employed without the consent (which shall not unreasonably be withheld) of the architect or supervising officer.

There are no provisions for nominating sub-contractors in respect of prime cost sum items, nor are there provisions for nominated suppliers. Instead, there are the new provisions for named sub-contractors.

Named sub-contractors can arise in two ways:

- By naming in the specification, schedules of work or contract bills;
- By naming in an instruction for the expenditure of a provisional sum included in the contract documents.

In each case the JCT Form of Tender and Agreement NAM/T must be used. It is expressly stated in the contract that the naming provisions do not apply to the execution of parts of the works by a local authority or by statutory authorities executing such work solely in pursuance of their statutory rights and obligations.

The total package of documents relating to named sub-contractors comprises the following:

(1)   JCT Form of Tender and Agreement NAM/T;
(2)   JCT Sub-Contract Conditions NAM/SC;
(3)   JCT Fluctuations Clauses and Formula Rules.

The JCT has also issued Practice Note IN/1 giving general guidance on the Intermediate Form and its use.

The JCT has not issued a separate agreement for exchange between the employer and the named sub-contractor to cover such matters as design responsibilities, but the RIBA and CASEC (the Committee of Associations of Specialist Engineering Contractors) have prepared such a document entitled RIBA/CASEC Form of Employer/Specialist Agreement – ESA/1. If an element of design is incorporated in the specialist works, it is strongly recommended that such an agreement is used.

Contrary to the established custom of specialist sub-contract costs being identified in main contract procedures, under the Intermediate Form the only costs of any relevance are the contractor's costs. Having received a full description of the work and the named person/firm involved (whether in the original documents or under a contract instruction), it is the contractor's responsibility to submit his price for the work allowing for all the relevant circumstances and the sub-contract quotation. The contractor is eventually paid this price and he then settles the sub-contract account within the terms of his sub-contract.

To enable this system to work properly the correct procedure must be followed and to simplify this the NAM/T Form of Tender and Agreement has been formulated.

The Form of Tender and Agreement is divided into three sections, as follows:

Section 1   Invitation to tender;
Section 2   Tender by sub-contractor;
Section 3   Articles of agreement.

The procedure for operating the naming provisions is as follows.

(1)   The architect prepares NAM/T by completing the whole of Section 1 with the details of the contract and sub-contract works together with indications of the sub-contract programme and the basis of fluctuations and sends the document to the sub-contractor(s) he wishes to tender.
(2)   The sub-contractor completes Section 2, quoting the sub-contract sum, the percentage additions for daywork, any additional attendances or other special requirements, gives the programme infor-

mation and details of fluctuations arrangements not already stated in Section 1, and returns the document to the architect.

At this point the architect must decide if he accepts the tender, or which one of a number of tenders he wishes to accept.

(3) The architect, having selected the tender he wishes to accept, incorporates the document as part of the contract tender documents.

There will then be a pause in the process while the contract tendering procedure is worked through. Assuming that a contract is signed, then

(4) The contractor and the sub-contractor complete Section 3 not later than 21 days after the contract is signed – according to the contract (clause 3.3.1) – either by signing or executing as a deed and the sub-contract is then effective. The conditions, NMA/SC, are incorporated automatically by reference in Section 3.

(5) The contractor notifies the architect of the date when he has entered into the sub-contract with the named person.

If, for some reason, the contractor is unable to enter into a sub-contract in accordance with the particulars given in the contract documents, he shall inform the architect, specifying the problem and the architect, being reasonably satisfied that this is justified, shall then either:

- Change the offending particulars; or
- Omit the work; or
- Omit the work and substitute a provisional sum.

Using this system, the sub-contractor will have available, at the time of tendering, all the information necessary to submit a proper price.

Nominated Suppliers

A supplier is nominated or deemed to be nominated if:

- A prime cost sum is included in the bills of quantities and if the supplier is named in the bills or is subsequently named by the architect; or
- A prime cost sum is included in an instruction regarding the expenditure of a provisional sum and if the supplier is named in the instruction, or is subsequently named by the architect; or
- In an instruction regarding the expenditure of a provisional sum or in a variation order the architect specifies materials or goods which can only be purchased from one supplier.

If the nomination of a supplier arises from the expenditure of a provisional sum or under a variation order, the materials or goods concerned must be made the subject of a prime cost sum.

If a supplier is named in the bills of quantities, but no prime cost sum is included, that supplier would not be a nominated supplier, it being left to the contractor to inform the architect prior to the contract being placed if he considers that the named supplier should be treated as a nominated supplier and a prime cost sum inserted.

It is advisable that decisions on nominated suppliers should be taken during the pre-contract period so that they can be dealt with in the proper manner in the bills of quantities.

The JCT now publish a form of tender for nominated suppliers and, although the use of this is not mandatory under the terms of the contract, it is strongly recommended that it be used as it includes an appendix which sets out in detail the terms which must appear in the nominated supplier's conditions of sale, these terms now being incorporated in JCT 80.

Obtaining Tenders

The provisions of JCT 80, both as regards nominated sub-contractors and nominated suppliers, and those of IFC 84 for named sub-contractors, though rather complicated, go a long way to eliminating many of the problems which have arisen in the past in this difficult contractual area. The requirements which are mandatory must, of course, be followed by the employer's professional advisers. Those which are optional should be carefully considered and it is advisable that the formal procedures which are available should be followed. The permissible short cuts should only be taken on very minor matters.

Use of the standard forms will also simplify obtaining competitive tenders, since it will ensure that the firms tendering do so on the precise terms required by the contract.

Standard forms of tender are not in themselves sufficient to ensure good tenders. The accompanying documentation must adequately cover the work involved. This will normally take the form of drawings together with a specification, or a performance specification and where appropriate, a sub-contract bill of quantities. In all these documents the high standards recommended for the main contract documents should be maintained.

Bills of quantities and specifications for sub-contract works should not include any prime cost sums. The architect has no power to issue instructions regarding the expenditure of prime cost sums in sub-contract documents, and if it is desired to nominate within a nominated sub-contract, the works, materials or goods concerned must be covered by a provisional sum.

Sub-Contractor Design

Neither JCT 80 nor IFC 84 contemplates design on the part of the contractor. Where such design is involved the JCT has special forms of contract which apply. They are the JCT 80 Standard Form With Quantities modified by the Contractor's Designed Portion Supplement and the JCT 81 Standard Form with Contractor's Design. These contracts, however, do not fall within the scope of this book since we have concentrated on the traditional system of consultancy design and contractors building to that design.

When the contract excludes contractor design, it necessarily follows that any sub-contract must do likewise and therein lies a common problem, for inevitably some nominated sub-contractors under JCT 80 and IFC 84 will be involved in elements of design.

Where such sub-contract design is properly planned and co-ordinated with the other production information available to the contractor, as it should be in terms of good practice, there is usually little practical problem. However, when failures occur, or when delay is experienced, the employer is liable to suffer considerable loss in terms of time and/or money and he will have no contractual right to recover his losses under either JCT 80 or IFC 84.

In recognition of this situation, a system of separate agreements has been incorporated in the sub-contract procedure, whereby a direct contractual link is formed between the employer and the sub-contractor. Those associated with JCT 80 and IFC 84 are discussed briefly below. However, as a matter of good practice, it is emphasized that whenever sub-contract design is incorporated into a contract which does not allow for it, protection for the employer along the lines indicated should always be recommended by the design team.

JCT 80 and NSC/W

Under the JCT 80 *1991 procedure* the Standard Form of Employer/ Nominated Sub-Contractor Agreement, 1991 Edition, NSC/W, is an integral part of the procedure for nomination.

Under NSCW, in consideration of nomination, the sub-contractor warrants to exercise all reasonable skill and care in:

- His design of the sub-contract works;
- His selection of materials and goods for the sub-contract works;
- Satisfying any performance specification requirements;

and will

- Supply all information and drawings in time (so as not to give cause for a claim for delay);
- Not default on his obligations under the sub-contract;
- Not cause delay (so as to give the main contractor causes to claim an extension of time);
- Indemnify the employer against loss arising from re-nomination if he defaults.

The employer warrants that:

- The architect will direct the contractor and inform the sub-contractor as to the amounts certified in interim and final certificates;
- He will operate the system of final payment of sub-contractors laid down in the main contracts;
- He will operate the provisions for direct payment laid down in the main contract.

The agreement also contains arbitration provisions. It is signed or exe- cuted as a deed by the parties.

In addition, it should be noted that the sub-contractor, after nomina- tion, has an obligation to agree the particular conditions in Part 3 of NSC/T with the contractor, and execute the Articles of Nominated Sub- Contract Agreement, NSC/A.

As noted earlier, provision is also made for the nominated sub-con- tractor to withdraw his tender with impunity, where he can demonstrate a reasonable objection to the identity of the contractor, when such infor- mation was not available at the time that he submitted his tender.

IFC 84 and ESA/1

The Form of Employer/Specialist Agreement drawn up by the RIBA and CASEC is not integrated with the naming provisions of the Intermediate Form, but is drafted specifically for use with that form.

The form is drafted with a number of alternatives so that it can be used in connection with a tender or an estimate (if sufficient information is not available) and for issue with the contract tender documents or under a contract instruction.

Under the form, the specialist warrants to exercise reasonable skill and care in:

- His design for the sub-contract works;
- His selection of materials and goods for the sub-contract works;
- Satisfying any performance specification requirements;

and will:

- Supply all information, including drawings, to the architect and contractor in due time;
- Perform the sub-contract works in due time.

There are no general notification or payment provisions in this agreement, as in NSC/W. However, there are optional provisions for the employer to be able to require the specialist to purchase materials and goods and to commence fabrications of components before the sub-contract is entered into between the contractor and sub-contractor and, if the sub-contract is never entered into, for the employer to pay for such materials and goods and components properly fabricated; whereupon they become the property of the employer.

In the case of an estimate being submitted, there is an extra provision that, in the event of no sub-contract resulting, the employer will pay for abortive design work.

A copy of the tender (NAM/T) or estimate, as appropriate, should be attached to the agreement before execution. When only signed, a £10 consideration becomes payable by the employer to the specialist. No consideration is necessary with a deed.

Chapter 8

Obtaining Tenders

Selective Tendering

In the introduction to this book reference was made to the changes which have taken place in procedures in the building industry:

- Changes necessitated by the use of specialized methods of construction.
- Where the contractor's participation is required in the design stage.
- By the use of larger factory-made components and building systems.
- To let the design and construction programmes overlap to meet the requirement by employer to reduce the pre-contract period.

All these circumstances involve the early appointment of the contractor and this calls for modification of traditional tendering procedures. This matter has been dealt with extensively in the Aqua book *Tenders and Contracts for Building*. However, traditional methods are still widely used, and set out in this chapter are good procedures for these cases. To some extent they can be modified in special cases. The principles to be followed are clearly and logically defined in the Code of Procedure for Single Stage Selective Tendering 1989, published by the National Joint Consultative Committee for Building (NJCC) in collaboration with the Department of the Environment and the Joint Committees for Scotland and Northern Ireland.

Open tendering should be avoided and tenders should normally be obtained from a selected list of contractors. This controls the number and suitability of contractors on the list and hence reduces wasted resources in abortive tendering. The list should be discussed and agreed with the employer about three months before tenders are required. The selected list may be drawn up from firms known to the architect, quantity surveyor, employer and by selection from firms who apply in response to advertisements in the technical and local press. These advertisements should indicate the size, nature and location of the job and the date the tender documents will be ready.

Under the rules of the European Economic Community, contracts for public works exceeding a certain threshold value must be advertised in the *Official Journal* of the European Communities before tenders are invited. It is still permissible to compile a selected list of contractors from those who respond to the advertisement and to that extent selective tendering in these circumstances is not affected. The rules are set out in EC Directive 71/305 (amended 1989).

The number of firms to be invited to tender should depend on the size and type of contract. The Code of Procedure for Single Stage Selective Tendering 1989 recommended that the number of tenderers should be limited to a maximum of six. For contracts up to £250 000, four or five tenderers are quite sufficient and for contracts of greater value, six should be the maximum.

The cost of tendering is high and the larger the tender lists become the greater will be the cost of abortive tendering and this must be reflected in building prices. When the list has been settled one or two names should be appended in order that they may replace any firms on the list that do not accept the invitation. The firms selected should be of similar size and standard and should be known to be suitable for the particular job. A check should be put on each firm's financial standing and record.

Where the employer has built up an approved list of contractors this should be reviewed on a regular basis to exclude firms that have performed unsatisfactorily or become unsuitable. Appropriate contractors will then be inserted into the list to replace those deleted.

The basic principle behind selective tendering is that any of the contractors could carry out the work properly in accordance with the contract. The final choice of contractor will then be based on the lowest bona fide tender.

Preliminary Enquiry

About a month before the tender documents are sent out, a letter should be sent to the selected contractors asking them whether they wish to tender. It is essential that as much relevant information as possible is set down in this preliminary enquiry. This will enable the contractor to decide whether he wishes to tender or not. The Code of Procedure for Single Stage Selective Tendering sets out standard letters of enquiry for various JCT contracts. Typical information to be supplied would be as follows.

(1) Building owner/employer.
(2) Consultants who have specified duties under the contract (e.g. architect and quantity surveyors).
(3) Other consultants (e.g. engineers and project managers).
(4) Location of site.
(5) Restrictions of site use.
(6) General description of works.

(7)   Approximate cost range.
(8)   Form of contract to be used.
    (a) Which clauses will/will not apply.
    (b) Contract under deed/under hand.
    (c) Collateral warranty requirements.
    (d) Liquidated damages.
(9)   Major known sub-contractors.
(10)   Elements of contractor's design.
(11)   Anticipated date of possession.
(12)   Indication of employer's requirements for pre- and post-contract programme for the works.
(13)   Value of bond as a percentage of contract sum (if required).
(14)   Approximate date for dispatch of all tender documents.
(15)   Date that tender is to be returned.
(16)   Examination and procedure for correction of priced bill.

At the same time, any contractors who asked to tender in reply to an advertisement and who were not included in the selected list should be so informed.

Tender Period

Contractors must be allowed ample time for tendering if realistic tenders are to be obtained. The time required by a contractor to prepare a tender depends not only on the size of the job, but also on its complexity. This can vary considerably. Contractor design elements must be given sufficient time for tendering and may be sent out in advance of the other tender documents. As a general rule, at least four weeks should be allowed for tendering, though in some cases for small simple works this might be reduced to three weeks, and in other cases more time will be needed. It is imperative that tenderers should have sufficient time to prepare sound and proper tenders including, where necessary, obtaining prices from specialist sub-contractors.

Tender Documents

The following documents should be sent to the selected tenderers.

- Two unbound copies of the bills of quantities .
- Two copies of the drawings in accordance with the requirements of the SMM.
- Two copies of the form of tender.
- An addressed envelope for the return of the tender suitably marked on both sides with the word 'Tender' and marked on the front with the name of the job and the time and date to be delivered.
- If the priced bills of quantities are to be returned at the time of the tender, an addressed envelope to contain the bills, marked with the name of the job and with a space for the contractor's name on the outside.

If it is proposed to use the method of dealing with errors in tenders as outlined in Alternative 2 of Section 6 of the Code of Procedure for Single Stage Selective Tendering, priced bills of quantities must be returned with the tender.

In the letter accompanying these documents the following information and instructions should be given.

- Date and time tenders are to be returned.
- Place where all other drawings may be inspected and with whom arrangements have to be made for the purchase of additional copies of special drawings if required by specialists.
- Arrangements for inspection of the site.
- Time and place of opening of tenders and whether contractors may be present.
- When priced bills of quantities are to be submitted at the time of tendering, assurances that these will not be opened unless the contractor's tender is under consideration and that priced bills will be returned to unsuccessful contractors.
- Method of correction of priced bills (Section 6 of the NJCC Code of Procedure for Single Stage Selective Tendering 1989) Alternative 1/Alternative 2 to apply.
- Instruction that contractors are to acknowledge receipt of all documents.

Receipt of Tender

When tenders are received they should in no circumstances be opened before the proper time.

Although it is customary to state that no pledge is given to accept the lowest or any tender, unless special circumstances arise, or if the tender is flawed or unacceptably qualified, the most advantageous tender should be accepted. The selection having been made, the quantity surveyor should, as quickly as possible, report on the following.

- The checking of the arithmetic in the priced bills of the prospective contractor and ensuring that any amendments notified during the tendering period have been made.
- Inspection of prices in the bills and advising whether they are fair and reasonable noting any abnormal prices.
- Checking that the tenderer's basic rates of materials, if any, are reasonable.
- Making any analysis of the tender considered necessary.
- Comparison with cost target and opinion on tender.
- Recommendation for acceptance or otherwise.

Correction of Errors in Tenders

In the event of a serious error of pricing or arithmetic in the tenderer's bills of quantities, this should be dealt with in accordance with the alternative stated in the formal invitation to tender.

Under Alternative 1 the tenderer should be given details of errors in the bills and afforded an opportunity of confirming or withdrawing this offer. If the tender withdraws, the priced bills of the second lowest should be examined. In Alternative 2 the tenderer should be given an opportunity of confirming his offer or of amending it to correct genuine errors. Should he elect to amend his offer and the revised tender is no longer the lowest, then professional judgement must be exercised in determining which other tenders are to be examined in detail.

Under either alternative, where the tenderer confirms his offer, an endorsement should be added to the priced bills indicating that all rates or prices (excluding preliminary items, contingencies, prime cost and provisional sums) inserted therein by the tenderer are to be considered as reduced or increased in the same proportion as the corrected total of priced items exceeds or falls short of the original total of such items.

Alternatively, there may be cases where it is appropriate to make the adjustment in the preliminaries, or profit if separately stated, and this is certainly a simpler method saving much calculation later. In either case the contractor's agreement must be obtained and the endorsement signed by both parties.

If under the second alternative the tenderer elects to amend his tender figure, and possibly some of the rates in his bills, then he should either be allowed access to his original tender to insert the correct details and to initial them, or be required to confirm all the alterations in a letter. If in the latter case his revised tender is eventually accepted, the letter should be conjoined with the acceptance, and the amended tender figure and the rates in it should be substituted for those in the original tender.

Report on Tenders and Acceptance

The architect as well as the quantity surveyor may wish to comment and report on the tenders received and both reports should go to the employer with a recommendation to assist him in making his final decision on the tender to be accepted.

Within the period of tender being open for acceptance, the employer should have either accepted a tender or rejected all tenders, or accepted a tender subject to modifications to be agreed.

A contract should be drawn up and signed as soon as possible and in any case before possession of the site is given. The date for possession should either be established in the tender documents, or at latest when the tender is accepted.

Notification of Tender Results

Once the contract has been let, each tenderer, including the successful one, should be supplied with a list of the tender prices. In this way the tenderer's actual figure is not disclosed while each tenderer knowing his own figure will see how he stood in relation to the rest.

The principal advantages of this established tendering procedure are that it is the most effective way of letting a contract on a competitive financial basis, and that at the same time it establishes for the employer an accurate indication of his financial commitment. However, both these advantages can largely be maintained when other methods of obtaining tenders are adopted.

Competition of varying degrees can be introduced in the early appointment of the contractor and financial control can be maintained by making full use of up-to-date cost control techniques.

There are variations in tendering methods and these are dealt with in the Aqua book *Tenders and Contracts for Building*, which covers the subject of the choice of appropriate tendering and contractual arrangements in various circumstances.

Bonds (Guarantee of Performance)

The employer may require the contractor to provide a bond for the due performance of the contract. Such a bond will normally be obtained by the contractor from an insurance company or a bank or alternatively, where the contractor is a subsidiary company of a larger organization, it may take the form of a guarantee from the parent company. The bond holder or guarantor and the terms of the bond must of course be approved by the employer and the amount of surety provided will normally be 10% of the contract sum. This amount would become available to the employer to meet any additional expense he incurs as a result of the contractor's failing to execute the contract or otherwise being in breach of his obligations under it. As the raising of a bond can be a financial burden on the contractor in terms of his borrowing, it is considered good practice that the bond should be released on the contract reaching practical completion.

Whether or not a bond is required is one of those matters which must be settled before the documents are sent out to tender so that the contractor who is responsible for all costs in connection with the bond, can include those costs in his price.

The Aqua book *Contract Administration* provides examples of performance bonds and parent company guarantees as contained in the NJCC Guidance Note 2.

Index